Lipid Metabolism

Lipid Metabolism

Edited by **Donna Thompson**

New York

Published by Callisto Reference,
106 Park Avenue, Suite 200,
New York, NY 10016, USA
www.callistoreference.com

Lipid Metabolism
Edited by Donna Thompson

International Standard Book Number: 978-1-63239-448-4 (Hardback)

Printed in the United States of America.

Contents

Permissions

List of Contributors

Preface

This book has been a concerted effort by a group of academicians, researchers and scientists, who have contributed their research works for the realization of the book. This book has materialized in the wake of emerging advancements and innovations in this field. Therefore, the need of the hour was to compile all the required researches and disseminate the knowledge to a broad spectrum of people comprising of students, researchers and specialists of the field.

Lipids (fats and oils) are an extensive range of organic molecules that activate several functions in organisms. Lipids are essential components of our diet, focusing on their important contribution in energy, representing 9 kcal/g (or 37.7 kJ/g), and by some components relevant to the metabolism, such as essential fatty acids, fat soluble vitamins and sterols (cholesterol and phytosterols). Besides this, lipids have vital roles in human growth and development, along with promoting, preventing and/or participating in the pathogenesis or eventually in the treatment of various diseases. This book demonstrates a comprehensive analysis of the structure and metabolism of lipids.

At the end of the preface, I would like to thank the authors for their brilliant chapters and the publisher for guiding us all-through the making of the book till its final stage. Also, I would like to thank my family for providing the support and encouragement throughout my academic career and research projects.

Editor

Introduction to Lipid Metabolism

Overview About Lipid Structure

Rodrigo Valenzuela B. and Alfonso Valenzuela B.

Additional information is available at the end of the chapter

1. Introduction

The term lipid is used to classify a large number of substances having very different physical - chemical characteristics, being its solubility in organic non-polar solvents the common property for their classification. Lipids are composed of carbon, hydrogen and oxygen atoms, and in some cases contain phosphorus, nitrogen, sulfur and other elements. In this context, fats and oils are the main exponents of lipids present in foods and in nutritional processes [1,2], being diverse fatty acids and cholesterol the most representative molecules due their important metabolic and nutritional functions [3,4]. The structural, metabolic and nutritional importance of lipids in the body is supported by numerous investigations in different biological models (cellular, animals and humans). Lipids have been instrumental in the evolution of species, having important role in the growth, development and maintenance of tissues [5,6]. A clear example of this importance is the elevated fatty acid concentration present in nerve tissue, especially very long-chain polyunsaturated fatty acids [7,8].

Fatty acids are, among lipids, of crucial relevance in the structure and physiology of the body because: i) forms an integral part of phospholipids in cell membranes; ii) are the primary source of energy (9 kcal /g or 37.62 kjoules/g); iii) in infants, provide more than 50% of the daily energy requirements; iv) some fatty acids are of essential character and are required for the synthesis of eicosanoids and docosanoids (of 20 and 22 carbon atoms, respectively), such as leukotrienes, prostaglandins, thromboxanes, prostacyclins, protectins and resolvins), and; v) some of them may act as second messengers and regulators of gene expression [9,10]. Besides fatty acids, cholesterol is another lipid that has important functions in the body, among which are: i) together with phospholipids is important in the formation of cell membranes; ii) constitutes the skeleton for the synthesis of steroid hormones (androgens and estrogens); iii) from its structure is derived the structure of vitamin D, and; iv) participates in the synthesis of the bile salts and the composition of bile secretion [11].

Lipids play a key role in the growth and development of the organism, where the requirements of these molecules (mainly fatty acids) will change depending on the age and physiological state of individuals [12]. Furthermore, lipids have crucial participation both, in the prevention and/or in the development of many diseases, especially chronic non-communicable diseases [13], affecting the lipid requirements in humans [14]. As food components, lipids are also important because: i) are significant in providing organoleptic characteristics (palatability, flavor, aroma and texture); ii) are vehicle for fat soluble vitamins, pigments or dyes and antioxidants, and; iii) may act as emulsifying agents and/or promote the stability of suspensions and emulsions [15].

Fats and oils, the most common lipids in food, are triacylglyceride mixtures, i.e. structures formed by the linking of three different or similar fatty acids to the tri-alcohol glycerol [16]. A fat is defined as a mixture of triacylglycerides which is solid or pasty at room temperature (usually 20 °C). Conversely, the term oil corresponds to a mixture of triglycerides which is liquid at room temperature. In addition to triacylglycerides, which are the main components of fats and oils (over 90%), these substances frequently contain, to a lesser extent, diacylglycerides, monoacylglycerides, phospholipids, sterols, terpenes, fatty alcohols, carotenoids, fat soluble vitamins, and many other minor chemical structures [17,18]. This chapter deals with the general aspects of lipids, especially those related to the chemical structure and function of these molecules.

2. Fatty acids

Fatty acids are hydrocarbon structures (containing carbon and hydrogen atoms) formed by four or more carbons attached to an acidic functional group called carboxyl group. The chemical and physical properties of the different fatty acids, such as their solubility in non-polar solvents and the melting point, will depend on the number of carbon atoms of the molecule. [19]. The higher the number of carbon atoms of the chain the higher will be melting point of the fatty acid. According to the chain length fatty acids are referred as short-chain fatty acids, those having four (C4) to ten (C10) carbons; as medium-chain fatty acids those having twelve (C12) to fourteen (C14) carbons; long-chain fatty acids to those of sixteen (C16) to eighteen carbons (C18); and very long-chain fatty acids those having twenty (C20) or more carbon atoms. Molecules having less than four carbon atoms (C2; acetic acid and C3; propionic acid) are not considered fatty acids due their high water solubility. On the other side, fatty acids of high number of carbon atoms are not frequent, however are present in significant amount in the brain of vertebrates, including mammals and human. In the human brain have been identified fatty acids as long as 36 carbon atoms [20].

The link between carbons in fatty acids, correspond to a covalent bond which may be single (saturated bond) or double (unsaturated bond). The number of unsaturated bonds in the same molecules can range from one to six double bonds. Thus, the more simple classification of the fatty acids, divided them in those that have not double bonds, named saturated fatty acids (SAFA), and fatty acids that have one or more double bonds, collectively named unsaturated fatty acids. In turn, when the molecule has one unsaturation it is classified as monounsaturated fatty acid (MUFA) and when has two to six unsaturations are classified as polyunsaturated fatty acids (PUFAs) [21]. The presence of unsaturation or double bonds in fatty acids is represented by denoting the number of carbons of the molecule followed by an indication of the number of double bonds, thus: C18:1 corresponds to a fatty acid of 18 carbons and one unsaturation, it will be a MUFA. C20: 4 correspond to a molecule having 20 carbons and four double bonds, being a PUFA. Now, it is necessary to identify the location of the unsaturations in the hydrocarbon chain both in MUFAs and PUFAs [22].

3. Nomenclature of fatty acids

According to the official chemical nomenclature established by IUPAC (International Union of Practical and Applied Chemistry) carbons of fatty acids should be numbered sequentially from the carboxylic carbon (C1) to the most extreme methylene carbon (Cn), and the position of a double bond should be indicated by the symbol delta (Δ), together to the number of the carbon where double bonds begins. According to this nomenclature: C18: 1, Δ9, indicates that the double bond is between carbon 9 and 10 [23]. However, in the cell the metabolic utilization of fatty acids occurs by the successive scission of two carbon atoms from the C1 to the Cn (mitochondrial or peroxisomal beta oxidation). This means that as the fatty acid is being metabolized (oxidized in beta position), the number of each carbon atom will change, creating a problem for the identification of the metabolic products formed as the oxidation progress. For this reason R. Holman, in 1958, proposed a new type of notation that is now widely used for the biochemical and nutritional identification of fatty acids [24]. This nomenclature lists the carbon enumeration from the other extreme of the fatty acid molecule. According to this notation, the C1 is the carbon farthest from the carboxyl group (called as terminal or end methylene carbon) which is designed as "n", "ω" or "omega". The latter notation is the most often used in nutrition and refers to the last letter of the Greek alphabet [25]. Thus, C18: 1 Δ9 coincidentally is C18: 1 ω-9 in the "ω" notation, but C18: 2 Δ9, Δ12, according to this nomenclature ω would be C18: 2 ω-6. What happens with fatty acids having more than a double bond? Double bonds are not randomly arranged in the fatty acid structure. Nature has been "ordained" as largely incorporate them in well-defined positions. Most frequently double bonds in PUFAs are separated by a methyl group (or most correctly methylene group) forming a -C=C-C-C=C- structure which is known as "unconjugated structure", which is the layout of double bonds in most naturally occurring PUFAs [26]. However, although much less frequently, there are also present "conjugated structures" where double bonds are not separated by a methylene group, forming a -C=C-C=C- structure. This particular structural disposition of double bonds, i.e conjugated structures, is now gaining much interest because some fatty acids having these structures show special nutritional properties, they are called "conjugated fatty acids". Most of them are derived from the unconjugated structure of linoleic acid (C18:2 ω-6) [27,28].

For the application of the "ω" nomenclature and considering the "order" of double bonds in unsaturated fatty acids having unconjugated stucture, it can be observed that by pointing the location of the first double bond, it will automatically determined the location of the subsequent double bonds [29]. Thus, C18: 1 ω-9, which has a single double bond at C9 counted from the methyl end, correspond to oleic acid (OA), which is the main exponent of the ω-9 family. Oleic acid is highly abundant both in vegetable and animal tissues. C18: 2 ω-6 corresponds to a fatty acid having double bonds at the C6 and C9 (for unconjugated fatty acids it is not necessary to indicate the position of the second or successive double bonds). This is linoleic acid (LA), the main exponent of the ω-6 family and which is very abundant in vegetable oils and to a lesser extent in animal fats [30]. C18: 3, ω-3 corresponds to a fatty acid having double bonds at C3, C6 and C9. It is alpha-linolenic acid (ALA), the leading exponent of the ω-3 family. ALA is a less abundant fatty acid, almost exclusively present in the vegetable kingdom and specifically in land-based plants [31]. Within (LCPUFAs), C20: 4, ω-6 or arachidonic acid (AA); C20: 5, ω-3 or eicosapentaenoic acid (EPA) and; C22 : 6, ω-3 or docosahexaenoic acid (DHA), are of great nutritional importance and are only found in ground animal tissues (AA) and in aquatic animal tissues (AA, EPA and DHA) and in plants of marine origin (EPA and DHA) [32].

The increase of double bonds in fatty acids significantly reduces its melting point. Thus, for a structure of the same number of carbon atoms, if it is saturated may give rise to a solid or semisolid product at room temperature, but if the same structure is unsaturated, may originate a liquid or less solid product at room temperature. Figure 1 shows the classification of fatty acids according to their degree of saturation and unsaturation and considering the notation "ω", and table 1 shows different fatty acids, showing the C nomenclature, their systematic name, their common name and the respective melting point.

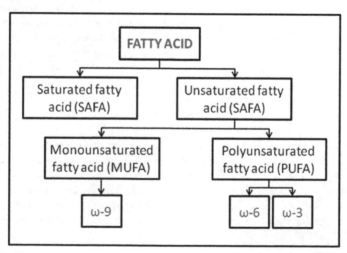

Figure 1. Classification of fatty acids according to their degree of saturation and unsaturation and considering the notation " ω".

Nomenclature	Systematic name	Common name	Melting point °C
Saturated Fatty Acid			
C 4:0	Butanoic	Butiric	-5.3
C 6:0	Hexaenoic	Caproic	-3.2
C 8:0	Octanoic	Caprilic	16.5
C10:0	Decanoic	Capric	31.6
C12:0	Dodecanoic	Lauric	44.8
C14:0	Tetradecanoic	Miristic	54.4
C16:0	Hexadecanoic	Palmitic	62.9
C18:0	Octadecanoic	Estearic	70.1
C20:0	Eicosanoic	Arachidic	76.1
C22:0	Docosanoic	Behenic	80.0
C24:0	Tetracosanoic	Lignocenic	84.2
Unsaturated Fatty Acid			
C16:1	9-Hexadececoic	Palmitoleic	0.0
C18:1	9-Octadecenoic	Oleic	16.3
C18:1	11-Octodecenoic	Vaccenic	39.5
C18:2	9,12-Octadecadienoic	Linoleic	-5.0
C18:3	9,12,15-Octadecatrienoic	Linolenic	-1.0
C20:4	5,8,11,14-Eicosatetraenoic	Arachidonic	49.5

Table 1. Different fatty acids, showing the C nomenclature, their systematic name, their common name and the respective melting point.

4. Mono-, di- and triacylglycerides

The structural organization of fatty acids in food and in the body is mainly determined by the binding to glycerol by ester linkages. The reaction of a hydroxyl group of glycerol, at any of its three groups, with a fatty acid gives rise to a monoacylglyceride. The linking of a second fatty acid, which may be similar or different from the existing fatty acid, gives rise to a diacylglyceride. If all three hydroxyl groups of glycerol are linked by fatty acids, then this will be a triacylglyceride [33]. Monoacylglycerides, by having free hydroxyl groups (two) are relatively polar and therefore partially soluble in water. Different monoacylglycerides linked to fatty acids of different lengths are used as emulsifiers in the food and pharmaceutical industry [34]. The less polar diacylglycerides which have only one free hydroxyl group are less polar than monoacylglycerides and less soluble in water. Finally, triacylglycerides, which lack of free hydroxyl groups are completely non-polar, but highly soluble in non-polar solvents, which are frequently used for their extraction from vegetable or animal tissues, because constitutes the energy reserve in these tissues [35]. Diacylglycerides and monoacylglycerides are important intermediates in the digestive and absorption process of fats and oils in animals. In turn, some of these molecules also perform other metabolic functions, such as diacylglycerides which may act as "second messengers" at the intracellular level and are also part of the composition of a new generation of oils nutritionally designed as "low calorie oils" [36]. When glycerol forms mono-, di-, or triacylglycerides, its carbon atoms are not chemically and structurally equivalent. Thus, carbon 1 of the glycerol is referred as carbon (α), or sn-1 (from "stereochemical number"); carbon 2 is referred as carbon (β), or sn-2, and carbon 3 as (γ), or sn-3. It is important to note that the notation "sn" is currently the most frequently used [37]. This spatial structure (or conformation) of mono-, di- and triacyglycerides is relevant in the digestive process of fats and oils (ref). Figure 2 shows the structure of a monoacylglceride, a diacylglyceride and a triacylglyceride, specifying the "sn-" notation.

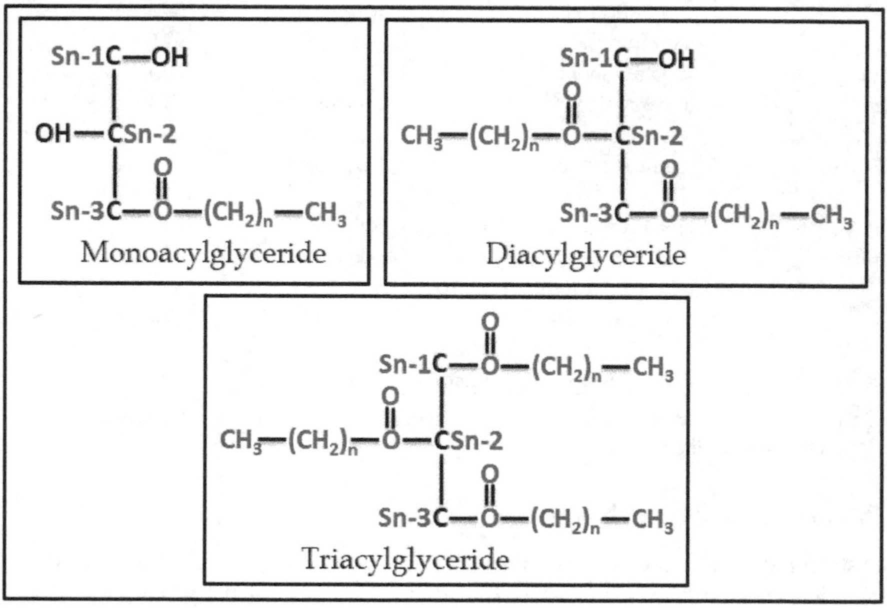

Figure 2. Structure of a monoacylglyceride, a diacylglyceride and a triacylglyceride, specifying the "sn-" notation

5. Essential fatty acids

The capability of an organism to metabolically introduce double bonds in certain positions of a fatty acid or the inability to do this, determines the existence of the so-called non-essential or essential fatty acids (EFAs). According to this capability, mammals, including primates and humans, can introduce a double bond only at the C9 position of a saturated fatty acid (according to "ω" nomenclature) and to other carbons nearest to the carboxyl group, but not at carbons nearest the C1 position [38]. This is the reason why OA is not an EFA. In contrast, mammals can not introduce double bonds at C6 and C3 positions, being the reason why AL and ALA are EFAs. By derivation, the AA is formed by the elongation and desaturation of LA, and EPA and DHA, which are formed by elongation and desaturation of ALA, become also essential for mammals when their respective precursors (LA and ALA, respectively) are nutritionally deficient [39]. Figure 3 shows the chemical structure of a SAFA, such as the stearic acid (C18:0), AO, LA and ALA, exemplifying the "ω" notation of each and indicating the essential condition in relation to the position of their unsaturated bonds.

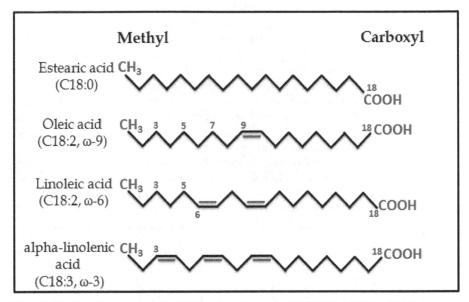

Figure 3. The chemical structure of a SAFA, such as the stearic acid (C18:0), AO, LA and ALA, exemplifying the "ω" notation of each and indicating the essential condition in relation to the position of their unsaturated bonds

6. Isomerism of fatty acids

According to the distribution of double bonds in a fatty acid and to its spatial structure, unsaturated fatty acids may have two types of isomerism: geometrical isomerism and positional isomerism. By isomerism it is referred to the existence two or more molecules having the same structural elements (atoms), the same chemical formula and combined in equal proportions, but having a different position or spatial distribution of some atoms in the molecule [40].

6.1. Geometrical isomers of fatty acids

Carbon atoms forming the structure of the fatty acids possess a three-dimensional spatial structure which forms a perfect tetrahedron. However, when two carbons having tetrahedral structure are joined together through a double bond, the spatial conformation of the double bond is modified adopting a flat or plane structure [41]. Rotation around single bonds (C-C) is entirely free, but when they are forming a double bond (C=C), this rotation is impeded and the hydrogen atoms that are linked to each carbon involved in the bond may be at the same side or opposed in the plane forming the double bond. If hydrogen atoms remain at the same side, the structure formed is referred as *cis* isomer (denoted as "*c*"). When hydrogen atoms remain at opposite sides the structure formed is referred as *trans* isomer (denoted as "*t*", *trans*: means crossed) [42]. Figure 4 shows the *cis* – *trans* geometric isomerism of fatty acids. The *cis* or *trans* isomerism of fatty acids confers them very different physical properties, being the melting point one of the most relevant [43]. Table 2 shows the melting point of various *cis* – *trans* geometric isomers of different fatty acids. It can be observed substantial differences in the melting point of *cis*- or *trans* isomers for the same fatty acid. Melting point differences bring to the geometrical isomers of a fatty acid very different biochemical and nutritional behavior. Fatty acids having *trans* isomerism, especially those of technological origin (such as generated during the partial hydrogenation of oils), have adverse effect on humans, particularly referred to the risk of cardiovascular diseases [44]. It is noteworthy that the majority of naturally occurring fatty acids have *cis* isomerism, although thermodynamically is more stable the *trans* than the *cis* isomerism, whereby under certain technological manipulations, such as the application of high temperature (frying process) or during the hydrogenation process applied for the manufacture of shortenings, *cis* isomers are easily transformed into *trans* isomers [45].

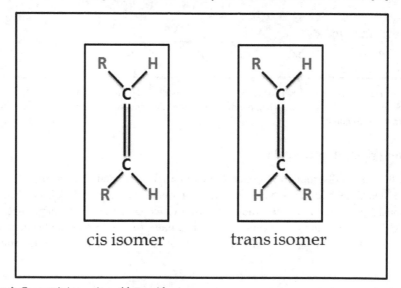

Figure 4. Geometric isomerism of fatty acids

6.2. Positional isomers of fatty acids

Positional isomerism refers to the different positions that can occupy one or more double bonds in the structure of a fatty acid. For example, oleic acid (C18:1 Δ9c), is a common fatty acid in vegetable oils, particularly in olive oil, but vaccenic acid (C18:1 Δ11t) is more common in animal fats. This is a double example, since both fatty acids are geometric isomers (oleic acid *cis* and vaccenic acid *trans*) and at the same time positional isomers, since oleic acid has a double bond at the $\Delta 9$ position and vaccenic acid at the $\Delta 11$ position [46].

Fatty acid	Isomerism	Melting point (°C)
C12:0	------	44.2
C16:0	------	62.7
C18:0	------	69.6
C18:1	Cis	13.2
C18:1	Trans	44.0
C18:2	cis, cis	-5.0
C18:2	trans, trans	18.5
C18:3	cis, cis, cis	11.0
C20:3	trans, trans, trans	29.5

Table 2. Changes in the melting point of various *cis – trans* geometrical isomers of different fatty acids

In general, all fatty acids naturally present positional isomerism of their more frequent molecular structure. However, these isomers occur in very low concentrations. Unlike the known biochemical and nutritional effects of *trans* geometric isomers, there is little information about the biological effects of positional isomers and for the majority of them these effects are considered as not relevant, except for some conjugated structures, such as conjugated linoleic acid (C18:2, $\Delta 9$, $\Delta 11$, CLA), a geometric and positional isomer of the most common linoleic acid, for which it has been attributed various health properties, especially those related to anti-inflammatory and lipolytic actions, but up to date the scientific evidence for these properties are considered insufficient [47]. Such as geometrical isomerism, the technological manipulation of fatty acids (i.e. temperature and/or hydrogenation) increases the number and complexity of the positional isomers [48]. Figure 5 summarizes the positional and geometric isomers of unsaturated fatty acids.

7. Phospholipids

Phospholipids are minor components in our diet because less than 4-5% of our fat intake corresponds to phospholipids. However, this does not detract nutritionally important to these lipids, since they are important constituents of the cellular structure having also relevant metabolic functions [49]. Life, in its origin, would not have been possible without the appearance of phospholipids, as these structures are the fundamental components of all cellular membranes. Phospholipids have structural and functional properties that distinguish them from their counterparts, triacylglycerides. In phospholipids positions sn-1 and sn-2 of the glycerol moiety are occupied by fatty acids, more frequently polyunsaturated fatty acids, linked to glycerol by ester bonds. The sn-3 position of glycerol is linked to orthophosphoric acid [50]. The structure which is formed, independent of the type of fatty acid that binds at sn-1 and sn-2, is called phosphatidic acid. The presence of phosphate substituent at the sn-3 position of the glycerol gives a great polarity to this part of the molecule, being non-polar the rest of the structure, such as in triacylglycerides. This double feature, a polar extreme and a non-polar domain due the presence of the two fatty acids characterizes phospholipids as amphipathic molecules (*amphi*: both; *pathos*: sensation) [51].

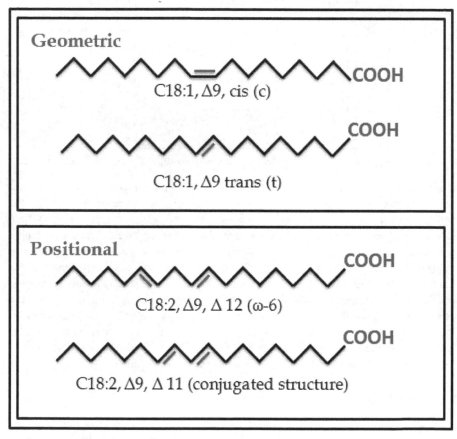

Figure 5. Positional and geometric isomers of unsaturated fatty acids

The structure of phospholipids is usually simplified representing the polar end as a sphere and the fatty acids as two parallel rods. Figure 6 shows the chemical structure of phosphatidic acid in its simplified representation. The amphipathic character of phosphatidic acid can be increased by joining to the phosphate different basic and polar molecules that increases the polarity to the extreme of the sn-3 position. When the substituent of the phosphate group is the aminoacid serine it is formed phosphatidylserine; when it is etanolamine it is formed phosphatidylethanolamine (frequently known as cephalin); when choline is the substituent it is formed phosphatidylcholine (well known as lecithin); and when the substituent is the polyalcohol inositol it is formed phosphatidylinositol, a very important molecule involved in cell signaling. [52].

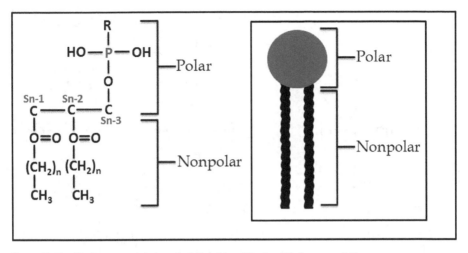

Figure 6. Chemical structure of phosphatidic acid and its simplified representation

These more complex phospholipids are much more common than phosphatidic acid, since this is only the structural precursor of the above molecules. Figure 7 shows the structure of various phospholipids. A number of other molecules are also classified as phospholipids, but are structurally different. Cardiolipin is a "double" phospholipid in which two phosphatidic acid molecules are attached through their phosphates by a molecule of glycerol. Cardiolipin is a very important in the structure of the inner membrane of mitochondria and due their molecular volume it is the only immunogenic phospholipid (which stimulates the formation of antibodies) [53]. Plasmalogens are other lipid molecules related to phospholipids. In these molecules the substituent at sn-1 position of the glycerol is not a fatty acid, but a fatty alcohol which is linked to glycerol by an ether linkage. Phosphatidalethanolamine (different than phosphatidylethanolamine) is an abundant plasmalogen in the nervous tissue [54]. Phosphatidalcholine, the plasmalogen related to phosphatidylcholine, is abundant in the heart muscle. Another structures related to phospholipids are sphingolipids. In these structures glycerol is replaced by the amino alcohol; sphingosine. When the hydroxyl group (alcoholic group) of sphingosine is substituted by phosphocholine, it is formed sphingomyelin, which is the only sphingolipid that is present in significant amount in human tissues as a constituent of myelin that forms nerve fibers [ref]. Platelet activating factor (PAF) is an unusual glycerophospholipid structure. In this molecule position sn-1 of glycerol is linked to a saturated alcohol through an ether bond (such as in plasmalogens) and at the sn-2 binds an acetyl group instead of a fatty acid. PAF is released by a variety of cells and by binding to membrane receptors produces aggregation and degranulation of platelets, has potent thrombotic and inflammatory effects, and is a mediator of anaphylactic reactions [55].

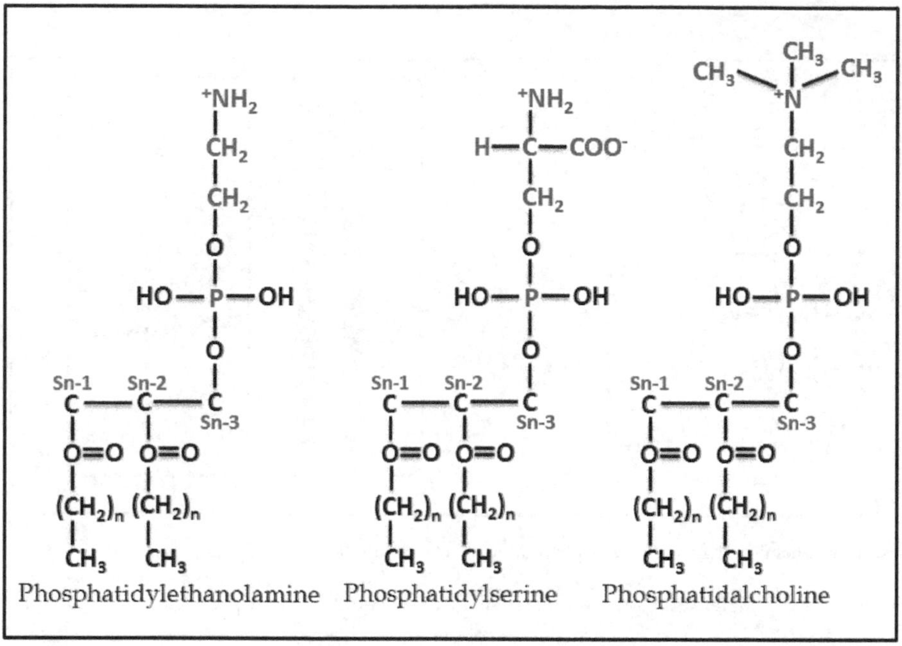

Figure 7. Structure of various phospholipids

A fundamental aspect of phospholipids is their participation in the structure of biological membranes, and the structural characteristics of the fatty acids are relevant to determine the behavior and the biological properties of the membrane. As an example, a diet rich in saturated fatty acids result in an increase in the levels of these fatty acids into cell membrane phospholipids, causing a significant decrease in both, membrane fluidity and in the ability of these structure to incorporate ion channels, receptors, enzymes, structural proteins, etc., effect which is associated to an increased cardiovascular risk [56]. By contrast, a diet rich in monounsaturated and/or polyunsaturated fatty acids produce an inverse effect. At the nutritional and metabolic level this effect is highly relevant because as the fatty acid composition of the diet is directly reflected into the fatty acid composition of phospholipids, changes in the composition of the diet, i.e. increasing the content of polyunsaturated fatty acids, will prevent the development of several diseases [57]. Figure 8 shows a simulation how the structural differences of the fatty acids which comprise phospholipids may affect the physical and chemical behavior of a membrane.

8. Sterols

Sterols are derived from a common structural precursor, the sterane or cyclopentanoperhydrophenanthrene, consisting in a main structure formed by four aromatic rings identified as A, B, C and D rings. All sterols have at carbon 3 of A ring a polar hydroxyl group being the rest of the structure non-polar, which gives them certain amphipathic character, such as phospholipids. Sterols have also a double bound at carbons 5 and 6 of ring B [58]. This double bond can be saturated (reduced) which leads to the formation of stanols, which together with plant sterols derivatives are currently used as hypocholesterolemic agents when incorporated into some functional foods. At carbon 17 (ring D) both sterols as stanols have attached an aliphatic group, consisting in a linear structure of 8, 9 or 10 carbon atoms, depending on whether the sterol is from animal origin (8 carbon atoms) or from vegetable origin (9 or 10 carbon atoms) [59]. Figure 9 shows the structure of cyclopentanoperhydrophenanthrene and cholesterol. Often sterols, and less frequent stanols, have esterified the hydroxyl group of carbon 3 (ring A) with a saturated fatty acid (usually palmitic; C16:0) or unsaturated fatty acid (most frequent oleic; C18:1 and less frequent linoleic acid; C18:2. The esterification of the hydroxyl group eliminates the anphipaticity of the molecule and converts it into a structure completely non-polar. Undoubtedly among sterols cholesterol is the most important because it is the precursor of important animal metabolic molecules, such as steroid hormones, bile salts, vitamin D, and oxysterols, which are oxidized derivatives of cholesterol formed by the thermal manipulation of cholesterol and that have been identified as regulators of the metabolism and homeostasis of cholesterol and sterols in general [60].

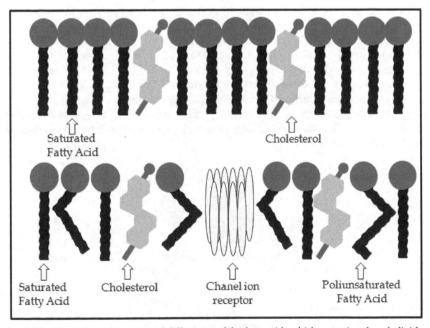

Figure 8. Simulation how the structural differences of the fatty acids which comprise phospholipids may affect the physical and chemical behavior of a membrane

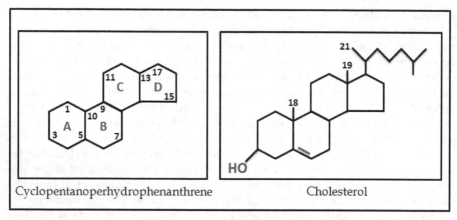

Figure 9. Structure of cyclopentanoperhydrophenanthrene and cholesterol

9. Conclusions

Lipids are a large and wide group of molecules that are present in all living organism and also in foods and characterized by particular physicochemical properties, such as their non polarity and their solubility in organic solvents. Some lipids, in particular fatty acids and sterols, are essential for animal and plant life. Lipids are key elements in the structure, biochemistry, physiology, and nutritional status of an individual, because are involved in: i) the cellular structure; ii) the cellular energy reserve, iii) the formation of regulatory metabolites, and; iv) in the regulation and gene expression, which directly affects the functioning of the body. Another important aspect related to lipids is their important involvement, either in the treatment and/or the origin of many diseases which can affect humans. Structural and functional characteristics of lipids, discussed in this chapter, will allow you to integrate those metabolic aspects of these important and essential molecules in close relationship of how foods containing these molecules can have a relevant influence in the health or illness of an individual.

Author details

Rodrigo Valenzuela B.
Nutrition and Dietetics School, Faculty of Medicine, University of Chile, Santiago Chile

Alfonso Valenzuela B.
Lipid Center, Nutrition and Food Technology Institute, University of Chile, Faculty of Medicine, University of Los Andes, Santiago Chile

Acknowledgement

The authors are grateful from FONDECYT, FONDEF and INNOVA-Chile the support of their research.

10. References

[1] Carpenter K. (1998). Early ideas on the nutritional significance of lipids, *J. Nutr.* 128, 423S-426S.

[2] Orešič M. (2009). Metabolomics, a novel tool for studies of nutrition, metabolism and lipid dysfunction. *Nutrition, Metabolism and Cardiovascular Diseases*, Vol 19, pp. 816-24.

[3] Burlingame B, Nishida C, Uauy R. et al. (2009). Fats and fatty acids in human nutrition; joint FAO/WHO Expert Consultation. *Ann. Nutr. Metab.*, Vol 55, pp. 1-3.

[4] Food and Agriculture Organization of the United Nations (FAO). (2010). Fats and fatty acids in human nutrition Report of an expert consultation. *Food and Nutrition.* Paper 91, pp. 9-19.

[5] Valenzuela A. (2009). Docosahexaenoic acid (DHA), an essential fatty acid for the proper functioning of neuronal cells: Their role in mood disorders. *Grasas & Aceites*, Vol 60, pp. 203-12.

[6] Simopoulos A. (2011), Evolutionary aspects of diet: the omega-6/omega-3 ratio and the brain. *Mol Neurobiol*, Vol 44, pp. 203-15.

[7] Valenzuela R, Bascuñán K, Valenzuela A, et al. (2009). Omega-3 fatty acids, neurodegenerative and psychiatric diseases: A new preventive and therapeutic approach. *Rev Chil Nutr*, Vol 36, pp. 1120-28.

[8] Campoy C, Escolano-Margarit M, Anjos T, et al. (2012). Omega 3 fatty acids on child growth, visual acuity and neurodevelopment. *Br J Nutr.* 2012, Vol 107, Suppl 2:S85-106.

[9] Weylandt K, Chiu C, Gomolka B, et al. (2012). Omega-3 fatty acids and their lipid mediators: towards an understanding of resolvin and protectinformation. *Prostaglandins Other Lipid Mediat.* Vol 97, pp. 73-82.

[10] Zúñiga J, Cancino M, Medina F, et al. (2011). N-3 PUFA supplementation triggers PPAR-α activation and PPAR-α/NF-κB interaction: anti-inflammatory implications in liver ischemia-reperfusion injury. *PLoS One.* Vol 6, pp. e28502.

[11] Dawson P, Lan T, and Rao A. (2009). Bile acid transporters *J. Lipid Res.* Vol 50, pp. 2340-57.

[12] Dietary fats and coronary heart disease. (2012). Willett WC. *J Intern Med.* Vol 272, pp. 13-24.

[13] Valenzuela R, and Videla L. (2011). The importance of the long-chain polyunsaturated fatty acid_n-6/n-3 ratio_in development of non-alcoholic fatty liver associated with obesity. *Food Funct.* Vol 2, pp. 644-8.

[14] Katan M, Zock P, and Mensink R. (1994). Effects of fats and fatty acids on blood lipids in humans: an overview. *Am. J. Clin. Nutr.*, Vol 60, pp. 1017S-22S.

[15] Valenzuela A, Delplanque B, and Tavella M. (2011). Stearic acid: a possible substitute for trans fatty acids from industrial origin. *Grasas & Aceites.* Vol 62, pp. 131-38.

[16] Lee Y, Tang T, and Lai O. (2012). Health Benefits, Enzymatic Production, and Application of Medium and Long-Chain Triacylglycerol(MLCT) in Food Industries: A Review. *J Food Sci.* doi: 10.1111/j.1750-3841.2012.02793.x

[17] Asensio C, Nepote V, and Grosso N. (2011) Chemical stability of extra-virgin_olive oil added with oregano essential oil. *J Food Sci.* Vol 76, pp. S445-50.

[18] Saggini A, Anogeianaki A, Angelucci D, et al. (2011). Cholesterol and vitamins: revisited study. *J Biol Regul Homeost Agents*. Vol 25, pp. 505-15.

[19] Fahy E, Subramanium S, Brown A, et al. (2005). A comprehensive classification system for lipids. *J. Lipid Res*. Vol 46, pp. 839-61.

[20] Kuipers R, Luxwolda M, Offringa P, et al. (2012). Fetal intrauterine whole body linoleic, arachidonic and docosahexaenoic acid contents and accretion rates. *Prostaglandins Leukot Essent Fatty Acids*. Vol 86, pp 13-20.

[21] Rodríguez E, Giri M, Rottiers R et al. Fatty acid composition of erythrocyte phospholipids is related to insulin levels, secretion and resistance in obese type 2 diabetics on Metformin. *Clinica Chimica Acta*, Vol 346. pp. 145-52.

[22] Gurr MI. (1992). Dietary lipids and coronary heart desease: Old evidence, new persspective *Prog in Lip Res*. Vol 31, pp 195-243.

[23] Martin D, Muriel E, Antequera T, et al. Quantitative changes in the fatty acid profile of lipid fractions of fresh loin from pigs as affected by dietary conjugated linoleic acid and monounsaturated fatty acids during refrigerated storage. *J. Food Composition and Analysis*. Vol 22, pp. 102-11.

[24] Arnauld S, Fidaleo M, Clémencet MC, et al. (2009). Modulation of the hepatic fatty acid pool in peroxisomal 3-ketoacyl-CoA thiolase B-null mice exposed to the selective PPARalpha agonist Wy14,643. *Biochimie*. Vol 91, pp. 1376-86.

[25] Harris W, Miller M, Tighe A, et al. (2008). Omega-3 fatty acids and coronary heart disease risk: Clinical and mechanistic perspectives *Atherosclerosis*, Vol 197, pp. 12-24.

[26] Osorio N, Ribeiro M, da Fonseca M, et al. (2008). Interesterification of fat blends rich in ω-3 polyunsaturated fatty acids catalysed by immobilized *Thermomyces lanuginosa* lipase under high pressure *J Mol Catalysis B: Enzymatic*, Vol 52. pp. 58-66.

[27] Skerratt J, Nichols P, Bowman J, et al. (1992). Occurrence and significance of long-chain (ω-1)-hydroxy fatty acids in methane-utilizing bacteria. *Organic Geochemistry*. Vol 18, pp. 189-94.

[28] Ansari G, Bhupendra S, Kaphalia M, et al. (1995). Fatty acid conjugates of xenobiotics *Toxicology Letters*. Vol 75, pp. 1-17.

[29] Michl J, and West R. Conformations of linear chains. (2000). Systematics and suggestions for nomenclature. Acc Chem Res. Vol 33, pp. 821-3.

[30] Din J, Newby D, and Flapan A, (2004). Omega 3 fatty acids and cardiovascular disease—fishing for a natural treatment. *BMJ*. Vol 3, pp. 30-35.

[31] Kris-Etherton P, Taylor D, Yu-Poth S, et al. (2000). Polyunsaturated fatty acids in the food chain in the United States. *Am J Clin Nutr. Vol 71. pp*. S179-88.

[32] Kris-Etherton P, Harris W, Appel LJ for the Nutrition Committee. (2002). AHA scientific statement. Fish consumption, fish oil, omega-3 fatty acids, and cardiovascular disease. *Circulation*. Vol 106. pp. 2747-57.

[33] Kishi T, Carvajal O, Tomoyori H, et al.(2002). Structured triglycerides containing medium-chain fatty acids and linoleic acid differently influence clearance rate in serum of triglycerides in rats *Nutrition Research*, Vol 22, pp. 1343-51.

[34] Carpentier Y, Simoens C, Siderova V, et al. (1997). Recent developments in lipid emulsions: relevance to intensive care. *Nutrition*, Vol 13, pp. 73-78.

[35] Irene Cetin, Gioia Alvino, Manuela Cardellicchio. (2009). Long chain fatty acids and dietary fats in fetal nutrition. *J Physiol*. Vol 15, pp. 3441–51.

[36] Rekha S, Singhal A, Gupta P, et al. Low-calorie fat substitutes. *Trends in Food Science & Technology*. Vol 2, pp. 241-44.

[37] Morita O, Knapp J, Tamaki Y, et al. (2008). Safety assessment of dietary diacylglycerol oil: A two-generation reproductive toxicity study in rats. *Food and Chemical Toxicology*, Vol 46, pp. 3059-68.

[38] Blasbalg T, Hibbeln J, Ramsden C, et al. (2011). Changes in consumption of omega-3 and omega-6 fatty acids in the United States during the 20th century. *Am J Clin Nutr*. Vol 93, pp. 950-62.

[39] Le H, Meisel J, de Meijer V, et al. (2012). The essentiality of arachidonic acid and docosahexaenoic acid. *Prostaglandins Leukot Essent Fatty Acids*. Vol 81, pp. 165-170.

[40] Valenzuela A, and Morgado N. (1999). Trans fatty acid isomers in human health and in the food industry. *Biol Res*. Vol 32, pp. 273-87.

[41] Micha R, King I, Lemaitre R, et al. (2010). Food sources of individual plasma phospholipid trans fatty acid isomers: the Cardiovascular Health Study. *Am J Clin Nutr*. Vol 91, pp. 883-93.

[42] Minville-Walz M, Gresti J, Pichon L, et al. (2012). Distinct regulation of stearoyl-CoA desaturase 1 gene expression by cis and trans C18:1 fatty acids in human aortic smooth muscle cells. *Genes Nutr*. Vol 7, pp. 209–16.

[43] Valenzuela A. (2008). Trans fatty acids I. Origin and effects in human health. *Rev Chil Nutr*. Vol 35, pp. 162-71.

[44] Micha R, and Mozaffarian D. (2008). Trans Fatty Acids: Effects on Cardiometabolic Health and Implications for Policy. Prostaglandins Leukot Essent Fatty Acids. Vol 79, pp. 147-52.

[45] Kuhnt K, Baehr M, Rohrer C, et al. (2011). Trans fatty acid isomers and the trans-9/trans-11 index in fat containing foods. Eur J Lipid Sci Technol. Vol 113, pp. 1281–92.

[46] Seppänen-Laakso T, Laakso I, Backlund P, et al. (1996). Elaidic and trans-vaccenic acids in plasma phospholipids as indicators of dietary intake of 18:1 trans-fatty acids. *Journal of Chromatography B: Biomedical Sciences and Applications*. Vol 687, pp. 996, Pages 371-78.

[47] Dilzer A, and Park Y. (2012). Implication of conjugated linoleic acid (CLA) in human health. Crit Rev Food Sci Nutr. Vol 52, pp. 488-513.

[48] Izadifar M. (2005). Neural network modeling of trans isomer formation and unsaturated fatty acid changes during vegetable oil hydrogenation. *Journal of Food Engineering*. Vol 66, pp. 227-32.

[49] Green J, Liu Z, and Bazinet P. (2010). Brain phospholipid arachidonic acid half-lives are not altered following 15 weeks of N-3 polyunsaturated fatty acid_adequate or deprived diet. *J Lipid Res*. Vol 51, pp. 535–43.

[50] Koga Y, and Goldfine H. (1984). Biosynthesis of_phospholipids_in Clostridium butyricum: kinetics of synthesis of plasmalogens and the glycerol acetal of ethanolamine plasmalogen. J Bacteriol. Vol 159, pp 597–604.

[51] García-Sáinz J, and Fain J. (1980). Effect of insulin, catecholamines and calcium ions on phospholipidmetabolism in isolated white fat-cells. *Biochem J*. Vol 15, pp. 781–89.

[52] Castro-Perez J, Roddy T, Nibbering N, et al. (2011). Localization of Fatty Acyl and Double Bond Positions in Phosphatidylcholines Using a Dual Stage CID Fragmentation Coupled with Ion Mobility Mass Spectrometry. *J Am Soc Mass Spectrom*. Vol 22, pp. 1552-67.

[53] Huang Z, Jiang J, Tyurin V, et al. (2009). Cardiolipin deficiency leads to decreased cardiolipin peroxidation and increased resistance of cells to apoptosis. *Free Radic Biol Med*. Vol 44, pp. 1935-44.

[54] Oleic- and Docosahexaenoic Acid-Containing Phosphatidylethanolamines Differentially Phase Separate from Sphingomyelin. Shaikh S, LoCascio D, Soni S, et al. (2010). *Biochim Biophys Acta*. Vol 1788, pp 2421-26.

[55] Esquenazi S, and Bazan H. (2010). Role of Platelet-Activating Factor in Cell Death Signaling in the Cornea: A Review. *Mol Neurobiol*. Vol 42, pp. 32-38.

[56] Gimenez M, Oliveros L, and Gomez N. (2011). Nutritional Deficiencies and Phospholipid Metabolism. *Int J Mol Sci*. Vol 12, pp 2408-33.

[57] Levantesi G, Silletta M, and Marchioli R. (2010). Uses and benefits of omega-3 ethyl esters in patients with cardiovascular disease. *J Multidiscip Healthc*. Vol 3, pp. 79-96.

[58] Gulati S, Liu Y, Munkacsi A, et al. (2011). Sterols and sphingolipids: Dynamic duo or partners in crime?. *Prog Lipid Res*. Vol 49, pp. 353-65.

[59] Izar M, Teganiv D, Kasmas S, et al. (2010). Phytosterols and phytosterolemia: gene–diet interactions *Genes Nutr*. Vol 6, pp. 17-26.

[60] Björkhem I. (2009). Are side-chain oxidized oxysterols regulators also in vivo?. *J Lipid Res*. Vol 50, pp. S213–18.

Molecular Aspects of Lipid Metabolism

Chapter 2

Oxidative Stress and Lipid Peroxidation – A Lipid Metabolism Dysfunction

Claudia Borza, Danina Muntean, Cristina Dehelean, Germaine Săvoiu, Corina Şerban, Georgeta Simu, Mihaiela Andoni, Marius Butur and Simona Drăgan

Additional information is available at the end of the chapter

1. Introduction

Free radicals are chemical compounds with unpaired electron(s), therefore being considered very active molecules. The cells had developed their own antioxidant defence systems in order to prevent the free radicals synthesis and to limit their toxic effects. These systems consist of enzymes which breakdown the peroxides, enzymes which bind transitional metals or compounds which are considered scavengers of the free radicals. Reactive species oxidize the biomolecules that will further elicit tissue injury and cell death. Evaluation of free radicals involvement in pathology is rather difficult due to their short life time.

2. Biochemistry of reactive oxygen species (ROS)

Free radicals can be formed by three mechanisms:

- Homolytic cleavage of a covalent bond of a molecule, each fragment retaining one electron

$$X{:}Y \rightarrow X^{\cdot} + Y^{\cdot}$$

- Loss by a molecule of a single electron

$$A \rightarrow A^{\cdot} + e^{-}$$

- Addition by a molecule of a single electron

$$A\,e^{-} \rightarrow A^{\cdot -}$$

- Heterolytic cleavage – covalent bond electrons are held up by only one of the molecule's fragments. Basically, charged ions occur.

$$X\,Y \;\rightarrow\; X\!\!:^- + Y^+$$

Oxygen activation is the main factor that induces enhanced formation of ROS. Due to its presence in the atmosphere, but also in the body, free radicals reaction with oxygen is inevitable. A second characteristic of oxygen refers to its electronic structure. Thus, O_2 has on the outer layer two unpaired electrons, each located on one orbital. Therefore, oxygen can be considered a free di-radical, but with a lower reactivity. Oxidation of this electron donor is achieved by spin inversion from the O_2 reaction with transition metals or by univalent reduction in two phases of one electron [5]. These two mechanisms underlie oxidation reactions that occur in nature. Although this process represents only 5%, following the univalent reduction of O_2, ROS occurs, with greater reactivity and toxicity, as is the hydroxyl radical OH.

In biological systems, the most important free radicals are oxygen derivate radicals formed by the following mechanisms:

O_2 reduction by the transfer of an electron will result in the synthesis of the superoxide anion (O_2^-). The formation of the superoxide anion is the first step of O_2 activation and occurs in the body during normal metabolic processes. In some cells, its production is continuous, which implies the existence of intracellular antioxidants [3].

$$O_2 + e^- \;\rightarrow\; O_{2^-}$$

Tissue alteration by traumatic, chemical or infectious means causes cell lysis along with the release of iron from deposits or by the action of hydrolases on metalloproteinase.

Reduced transition metal autooxidation generates the superoxide anion. The reaction of transition metal ions with O_2 can be considered a reversible redox reaction, important in promoting ROS formation.

$$Fe^2 + O_2 \;\rightarrow\; Fe^3 + O_{2^-}$$

$$Cu + O_2 \;\rightarrow\; Cu^2 + O_{2^-}$$

Degradation of H_2O_2 in the presence of transition metal ions leads to the formation of the most reactive and toxic ROS: the hydroxyl radical (OH.) (Fenton and Haber-Weiss reaction). To this radical, the body does not present antioxidant defense systems such as for the superoxide anion or hydrogen peroxide (H_2O_2). Although metallothioneins (natural antioxidants) are proteins that bind to metal ions, including Fe^{2+}, thus inhibiting the Haber-Weiss reaction, however they are found in too low concentrations in the body to be effective in the decomposition of the hydroxyl radical. But these reactions can be inhibited by specific scavengers for OH, such as mannitol and chelating agents: desferroxamine. However, chelators as EDTA stimulate this reaction.

The reduction of O_2 by two electrons leads to the formation of hydrogen peroxide, H_2O_2.

$$O_2 + 2e^- + 2H \rightarrow H_2O_2$$

H_2O_2 is often formed in biological systems via peroxide anion production.

$$2O_{2-} + 2H \rightarrow H_2O_2 + O_2$$

H_2O_2 is not a free radical, but falls within the category of reactive oxygen species that include not only free radicals but also its non-radical derivatives involved in producing these ROS. Of all free radicals, H_2O_2 is the most stable and the easiest to quantify. Intracellular formation of hydrogen peroxide, depending on the content of catalase, is the way by which the bactericidal mechanism is achieved in phagocytosis.

The formation of singlet oxygen ($^1O_2$). It represents an excited form of molecular oxygen, resulting from the absorption of an energy quantum. It is equated to a ROS due to its strong reactivity. Singlet oxygen has an electrophilic character, reacting with many organic compounds: polyunsaturated fatty acids, cholesterol, hydroperoxides or organic compounds containing S or N atoms, producing oxides. In plasma it is neutralized by the presence of antioxidants, especially albumin.

Singlet oxygen is formed in the following reactions:

- Reaction of hydrogen peroxide or hydroxyl radical with the superoxide anion
- Different enzymatic catalyzed reactions
- Decomposition of endoperoxides
- Degradation of hydroperoxides in liver microsomes

2.1. Free radicals resulting in lipid peroxidation propagation phase

Lipid peroxidation is a complex process consisting of three major phases: initiation, propagation and end of the reaction. The initiation phase is slow due to the need of accumulation of a sufficient quantity of ROS, followed by the activation process of oxygen which is the amplifier factor. The process' latency period is that which determines the continuation of reactions by altering the oxidative balance in favor of pro-oxidant factors. The evolution of these reactions is unpredictable due to the formation of own catalysts determining the complexity of the process [1].

Free radicals are very unstable, their lifetime being very short. Their reactivity results from their coupling at the end of the reaction, only for an unpaired electron to reappear, thus stimulate the propagation of the reaction by forming a new radical.

The end of the reaction occurs by:

- Free radical recombination among them or,
- Intervention of antioxidant systems with membrane or intracellular action: superoxide dismutase (SOD), catalase.

Peroxides and their decomposition products (aldehydes, lipofuscin) are the most stable and represent the final link of O_2 activation. They are produced directly by the hydroxyl or singlet oxygen radical. During these reactions, own catalysts are formed, represented by free radicals or degradation products that diversify and increase the oxidation reactions; the structures involved are diverse, and are represented by polyunsaturated fatty acids, hemoproteins, nucleic acids, carbohydrates or steroids [4].

3. The production of reactive oxygen species

3.1. Endogenous production

The electron transfer in the respiratory chain involves an incomplete reduction of molecular O_2 at a rate of 1-2% with the formation of superoxide anion and of singlet oxygen.

If the anion is released in a low in protons environment, it will initiate peroxidation, the substrate being formed by polyunsaturated fatty acids from cell membranes.

If the anion will reach a proton rich environment, dismutation will take place; this following auditioning an electron from another anion and by proton reaction will form hydrogen peroxide. Dismutation can occur spontaneously, but in this case it takes place very slowly or catalyzed by SOD, which increases 10^{10} the reaction rate to the body's pH. There is an inversely proportional relationship between reaction rate and pH value. The efficiency of this enzyme is proven by its presence in all aerobic cells, and cells exposed to oxygen action, as hepatocytes and erythrocytes, contain large amounts of SOD [6].

$$O_2 + e^- \rightarrow O_2^- + 2H \rightarrow H_2O_2$$

Superoxide anion production during mitochondrial respiration has a self-regulation mechanism. Superoxide anion formed in part by autoxidation of NADH dehydrogenase, can then induce this enzyme's inactivation, so the presence of SOD in the membrane matrix to achieve dismutation is absolutely necessary. It results that the two enzymes SOD and NADPH dehydrogenase are a metabolic control and energy preservation couple in the presence of oxygen [7].

The release of hydrogen peroxide is proportional to the partial pressure of O_2. In case of a cerebral or cardiac ischemia, extramitochondrial concentration decreases, disrupting oxidative phosphorylation and ATP levels. An inversely proportional relationship between mitochondrial H_2O_2 formation rate and lifetime exists. Thus, it was observed experimentally that old animals present an increase in lipid peroxide formation in the mitochondria as a result of increased production of superoxide anion compared with young animals.

Antimicrobial defense. Phagocytosis of bacterial germs is accompanied by a massive production of superoxide anions and other derivatives (OH^-, $HOCl$, H_2O_2, $1O_2$) from the leukocyte metabolism. A NADPH-dependent oxidase, activated by protein kinase C and arachidonic acid released under the action of phospholipase A allow anion synthesis with an increased consumption of O_2.

The sequence of reactions initiated in the membrane continues into the cytoplasm where a substantial amount of superoxide anion is formed which then is diffuses also extracellularly. Increased use of glucose occurs for energetic purposes and for restoring NADPH and oxygen consumption necessary for the production of ROS [8].

Hydrogen peroxide is toxic on the neutrophil, which is inhibited by the presence at this level of the three enzymes that degrade the excess of peroxide: GSH-peroxidase, catalase and myeloperoxidase.

The enzyme present in phagosome, myeloperoxidase, will catalyze in the presence of H_2O_2 and chloride ions, forming toxic halogenated derivatives.

$$H_2O_2 + Cl^- \rightarrow ClO^- + H_2O$$

In turn, hypochlorous acid can react with aminic groups or with the ammonium ion (NH_4) forming chloramines. In the presence of hydrogen peroxide, HOCl forms singlet oxygen. These products of activated leukocytes have bactericidal properties.

Based on the properties of leukocytes to emit chemiluminescence during phagocytosis, this method has a clinic utility. Chemiluminescence emission is due to formation of free radicals, lipid peroxides and prostaglandin synthesis, a process associated with phagocytosis. This property is suppressed by anesthesia, cytostatic agents and anti-inflammatory preparations. Drugs with anti-inflammatory effect inhibit the activity of cyclooxygenase, the enzyme involved in prostaglandin synthesis.

A deficiency in the leukocyte production of free radicals (septic granulomatosis) or decrease of myeloperoxidase activity (following corticotherapy) is characterized by particularly sensitivity to infections.

During phagocytosis, three cytotoxic and antimicrobial effect mechanisms take place:

- oxygen dependent mechanism involves activation of myeloperoxidase and other peroxidases
- Nitrogen compounds dependent mechanism involving participation of NO, NO_2, other nitrogen oxides and nitrites. In this mechanism both types of cytotoxic inorganic oxidants interact: oxygen and nitrogen reactive radicals.
- The third mechanism is independent of oxygen and nitrogen by changing phagolysosome pH that favors the action of antimicrobial substances present in the lysosomal or nuclear level.

The constitutive form of NO synthase is found in endothelial cells, neutrophils, neurons. The existence of the inducible form has been shown in macrophages, hepatocytes, endothelial cells, neutrophils and platelets. Glucocorticoids inhibit the expression of inducible NO synthase but not of the constitutive enzyme.

Nitrogen reactive radicals have a cytotoxic effect by inhibiting mitochondrial respiration, DNA synthesis, and mediate oxidation of protein and non-protein sulfhydryl groups.

Although NO has a protective role at the vascular level by a relaxing effect (EDRF), under certain conditions it may exert a cytotoxic effect, causing pathological vasodilatation, tissue destructions, inhibits platelet aggregation, modulates lymphocyte and immune response function.

Synthesis of prostaglandins. Phospholipase A_2 catalyzes the degradation of membrane phospholipids with arachidonic acid formation. Stimuli such as phagocytosis, antibody production, and immune complex formation, the action of bacterial endotoxins or cytokines stimulate the activation of this enzyme. There are two enzymatic forms: type I PLA_2, membrane bound, which is stimulated by Ca^{2+} at physiological pH, and the type II one, cytoplasmic, which is inhibited by Ca^{2+} and is active at acidic pH. Two enzymes, lipoxygenase and cyclooxygenase, bound to plasmic and microsomal membranes, convert arachidonic acid in derivatives such as: thromboxane, prostaglandins, leukotrienes [18].

Under the action of lipoxygenase, arachidonic acid is converted into a hydroperoxide: hydroperoxyeicosatetraenoic acid (HPETE) which will release the hydroxyl radical during its transformation into hydroxyeicosatetraenoic acid (HETE). Hill et al. have emphasized the role of glutathione peroxidase (GSH-Px) and of glutathione in this reaction: blocking the activity of this enzyme, they have noticed a significant decrease (of 66%) of HPETE conversion in HETE [14, 16].

Under the action of cyclooxygenase, arachidonic acid incorporates two oxygen molecules to form an endoperoxide, PGG; it loses the OH group to form PGH. This transformation, which is accompanied by the release of hydroxyl radical, exerts a negative retro-control to prostaglandin synthesis, inactivating the cyclooxygenase. Some of the products developed have a complex effect on the inflammatory process: thus, in the first phase, PGE_2 acts on cells from the vascular wall with a procoagulant effect, and in the late phase it has an inflammatory effect by inhibiting leukocyte activation and oxidative metabolism of these cells during phagocytosis. The byproducts resulting from this process will be the ones to modulate the intensity of the next phase [15, 22].

The two endoperoxides formed, PGG_2 and PGH_2, have an inducible role on the production of PCI_2 or TxA, being involved in the mechanism that ensures homeostasis of the vascular and platelet phase of hemostasis.

The other enzyme has a dual effect, and promotes the initiation of lipid peroxidation and the decomposition of resulting products of these reactions.

3.2. Exogenous production

The human body is subjected to aggression from various agents capable of producing free radicals. Thus, UVs induce the synthesis of ROS and free radicals generating molecules via photosensitizing agents.

Ingestion of alcohol causes ROS synthesis by different mechanisms: xanthine oxidase and aldehyde oxidase can oxidase the main metabolite of ethanol (acetaldehyde) resulting in superoxide anion.

Ethanol also stimulates the production of superoxide anion and, by inducing NADPH-oxidase synthesis, NADPH cytochrome reductase and P450 cytochrome.

The alcohol ingestion decreases the activity of protective enzymes (SOD, glutathione peroxidase). Also low serum concentrations of selenium and vitamin E have been found in alcoholics.

Toxic substances as nitrogen oxide and nitrogen dioxide in the environment are responsible for autoxidation of polyunsaturated fatty acids in lung alveoli. The reaction may be reversible or irreversible. $NO^.$ and NO_2 may react with H_2O_2 produced by alveolar macrophages and can generate the hydroxyl radical.

The reduction of carbon tetrachloride (CCl_4) in $CCl_3^.$ performed under the action of cytochrome P_{450} or in the presence of Fe_2 is another factor that induces autoxidation of polyunsaturated fatty acids, increasing lipid hydroperoxides concentration.

Anticancer drugs are able to synthesize free radicals, this process depending on the mode of action and their toxicity.

These drugs under the action of cytochrome P450-dependent enzymes produce the activation of O_2 with the formation of ROS which will attack GSH and other thiols (hemoglobin), causing the formation of lipid peroxides and activation of Ca^{2+}-dependent endonucleases.

These mechanisms can induce disturbances of the coagulation system (increased hemolysis), severe forms of cardiomyopathies, because of the low level of cardiac antioxidants (AO) [25].

4. Reactive oxygen species and oxidative stress

In physiological conditions a delicate balance exists between ROS production and the antioxidant capacity. A higher ROS production and/or a decreased antioxidant capacity is responsible for the harmful effects of free radicals or the oxidative stress (OS). Oxidative stress represents an important pathogenic mechanism involved in inflammation, cancerogenesis or aging [24].

End products of free radicals action, aldehydes, inhibit the activity of membrane enzymes (glucose-6-phosphate, adenylate cyclase). These aldehydes react selectively with proteins or enzymes containing SH groups and cause tissue destructions.

The emergence of OS is one of the most important pathogenic mechanisms involved in inflammation, carcinogenesis, radiation disease and aging.

The objective of various experimental models was to study erythrocyte response to oxidant substances action. Erythrocyte characteristics and test substance dosing allowed the evaluation of OS; these experiments can be extrapolated to explain various physiological or pathological processes in the body.

Oxidative stress is an ongoing process in the body, and under physiologic conditions there are effective mechanisms that negate its effects, thus high concentration of erythrocyte GSH and related enzyme equipment provide a defense against ROS.

Erythrocyte congenital enzyme deficiencies confer erythrocytes an increased sensitivity to OS.

A section of the body intensely studied to assess OS is the liver due to its role in the metabolism of a wide range of endogenous and exogenous products. Thus, by the metabolism of aromatic compounds, drugs or carcinogenic hydrocarbons in the live, a large amount of FR occurs, which will initiate in the next phase OS from this level.

Liver antioxidant systems are represented by SOD, GSH and dependent enzymes (transferase and peroxidase). Using ESR and spin trapping, FR resulting from chemical pollutants metabolism were identified, and a strong correlation between the functional impairment of the hepatic parenchyma, free radicals formation and decrease in GSH was noted. Under these conditions, free radicals of that substance occur which can cause tissue destructions also without O_2 activation.

The experimental poisoning of rats with alcohol (1.5 mmol/kg) showed significant decrease at one hour of ingestion of GSH, vitamin E and C along with hepatic necrosis and formation of lipid peroxides [2, 11].

GSH is an important protective factor against OS. Its level is interrelated with other antioxidants (vitamin C and E) that stimulate its preservation in reduced form.

Stress proteins. Structurally altered intracellular protein group, whose synthesis is induced by oxidative stress has been named stress proteins group. Of particular interest is the 32 kDa protein whose synthesis is induced by the action of ionizing radiation, hydrogen peroxide. This protein is a marker of generalized response to oxidative stress. Free radicals affect cytokines (endogenous pyrogens: IL-1, IL-2, TNF-a, IFN) that play a role in regulating signal transmission in response to stress which will cause the synthesis of these proteins. It is considered that SOD itself is a stress protein.

The tissue repair process is enzymatically catalyzed (repair enzymes) that break down damaged cellular particles, take intact aminoacids to synthesize new defense proteins.

5. The effects of reactive oxygen species

5.1. Oxygen free radicals – Intracellular messengers

Before discussing the negative effect of oxygen activation on the body, we should also take into consideration their involvement in certain physiological processes, when these ROS are produced in quantities which do not exceed the antioxidant capacity. Thus, the superoxide anion is produced by leukocytes during phagocytosis and in smooth muscle cells, epithelial cells, skin fibroblasts and endothelial cells [27].

The anion produced by macrophages and endothelial cells induces conformational changes of receptors on the LDL lipoprotein surface, allowing their recognition and involvement in atherogenesis [10, 12].

It is also involved in cascade-type metabolic reactions of arachidonic acid and in achieving platelet adhesion and aggregation function.

5.2. Lipid peroxidation

Formation of peroxides, especially lipid ones, is a consequence of the activation of O_2, the interconversion of reactive species and natural systems protection overcoming. In biological environments, the most favorable substrate for peroxidation is represented by polyunsaturated fatty acids (PUFA), components of cell and subcellular membranes.

Peroxidation is a complex process that includes three phases: initiation, propagation, end-decomposition, which interpose, so that only end products can be determined chemically: aldehydes (malondialdehyde), polymerized carbonyl compounds (lipofuscin) [9].

A radical character initiator (which may have different structures and origins, including peroxy ROO· radicals) removes a hydrogen atom from polyunsaturated fatty acid diallyl carbon, forming a favorable reactive center for oxygen action. The peroxy ROO· radicals which become hydroperoxides result. In fact, due to side reactions, other locations of the peroxide group per PUFA molecule occur [28].

5.3. Cell structural alterations

Since the formation of peroxides and their decomposition products, the sequence of reactions passes from a molecular level to a cellular one due to structural changes that occur in membranes: structural disorganization of the membrane and deterioration of pores crossing the double phospholipid layers. Peroxidation leads to changes in fatty acid qualitative composition of phospholipids composition with changing the ratio between PUFA and other acids. The first two effects induce the third, which consists in a decrease in membrane fluidity and altered active ion transport; these effects finally lead to changes in ion and other intracellular compounds concentration [26].

Numerous experimental studies have shown that tissue injury caused by free radicals determined at one point an imbalance of Ca^{2+} (i.e., increases in intracellular Ca^{2+} concentration). Under physiological conditions, there are effective homeostatic mechanisms (enzyme systems, protein transporters) to keep an optimum ratio between intracellular (0.1-0.4 microM) and extracellular of the mM order concentration. Overcoming these mechanisms (in this case by producing free radicals) determines the accumulation of calcium in the cell which will lead to structural membrane alterations, production of unsaturated lipids, efflux of GSH, its transition to an oxidized form and the creation of an intracellular oxidative potential [21, 23].

Experimental studies on isolated hepatocytes have shown the correlation between the cellular toxicity of calcium and the decrease of tocopherols levels, substances with strong antioxidant character.

Maintaining the cell functional state ultimately depends on the level of proteins containing SH groups. Thus, the role of GSH in protection against oxidative stress is precisely regeneration of protein SH groups which in turn will ensure intracellular calcium homeostasis. Vitamin E stabilizes ATPase activity dependent of calcium in the endoplasmic reticulum by maintaining SH groups in the structure of the enzyme in reduced state. Also, vitamin E is protective against the compounds resulting from lipid peroxidation: a molecule of alpha-tocopherol protects against 500 molecules of polyunsaturated fatty acids.

5.4. DNA destruction

The results of chromatographic technique used to determine the urinary excretion products resulting from scission of DNA in humans showed a normal excretion in average of 100 nmol products. This total represents 10^3 thymine molecules oxidized per day for each of the 6×10^{13} cells in the body.

Between eliminating these products and the specific metabolic rate (SMR) there is a linear correlation.

The specific metabolic rate of an organism is dependent on the O_2 use rate by its tissues and it is proportional to the free radicals production rate. In this case, the ratio between the total concentration of antioxidants (enzymatic and non-enzymatic systems) and the metabolic rate represents the protection degree of a tissue or body to free radicals. It seems that there is a genetic programming of the metabolic rate for each species and individual.

Looking at the hypothesis on free radicals involvement in aging, it has been shown that there is an inversely proportional relationship between the metabolic rate, free radicals production, respectively, and he maximum lifespan potential (MLP). Thus, on the evolutionary scale, metabolic rate decreased and lifespan increased, in mammals their product being constant.

One can calculate the lifespan energy potential (LEP), expressed in kcal-kg as follows: LEP 2.70 X MLP X SMR. This potential is directly proportional to the total concentration of antioxidants.

During aging, the formation of free radicals amplifies by exposure to prooxidant factors from the environment, and by the decreased antioxidant defense capacity.

At an intracellular level (especially in muscles and neurons), deposits of lipofuscin pigments, lipid peroxides and their breakdown products are formed.

These deposits are mainly localized in the myocardium, brain, and, by the age of 80, they represent 70% of cytoplasmic volume of neurons and 6% of that of myocardiocytes.

Experimental studies demonstrated that in 50 years a person accumulates 13.4 mg/lipofuscin/gram of myocardium, pigment formation taking place once with exceeding the absorption of 0.6 free radicals micromoles/gram of tissue.

There is an inversely proportional relationship between the formation of these products and the concentration of vitamin E in the body.

To control the effects of aging, ones requires a moderate diet, which reduces metabolic rate and O_2 consumption with an optimal concentration of lipids and a quantitatively and qualitatively balanced intake of antioxidants and other factors that enhance assimilation and their metabolism. It is also necessary to achieve a balanced interaction of endogenous antioxidants.

The antioxidants level varies greatly depending on the age of the body, that organ and subcellular components; thus an increase of GSH-Px activity was noted in mitochondria of cardiac cells and erythrocytes in the elderly, and a decrease of activity in liver and kidneys. The decrease of SOD activity in the liver of the elderly was highlighted and no significant changes in the concentration of intramitochondrial SOD in the heart were noted.

Also, there is a correlation between the intensity of DNA destructions caused by FR and xanthine oxidase concentration. This enzyme, present in low concentrations, in tissue or plasma, increases under tissue injury.

5.5. Effects on molecules

Free radicals are responsible for the inactivation of enzymes especially of serine proteases, the fragmentation of macromolecules (collagen, proteoglycans, hyaluronic acid), the formation of dimers, the protein aggregates in the cytoplasmic membranes. The most susceptible amino acids to their action are tryptophan, tyrosine, phenylalanine, methionine and cysteine.

Transition metal ions (Fe, Cu, Ni, Co, Cd) have a pro-oxidant action by intensifying reactions in which FR are formed and those in which the decomposition of lipid peroxides takes place. At the molecular level, Fe^{2+} ion contributes to the induction of oxidative stress by increasing non-enzymatic oxidation of catecholamines and GSH, promoting lipid peroxide decomposition and the formation of the most toxic free radicals, the hydroxyl radical. Fe^{2+}, under complexed form as transferrin, is inactive against peroxides. Fe^{2+} release from transferrin takes place under pH decrease as it does in hypoxia, leukocyte activation or in muscle tissue during strenuous physical exercise. Another source of free Fe^{2+} is represented by hemoglobin, which at low concentrations acts as a pro-oxidant favoring PUFA peroxidation. Proteins that bind Fe^{2+} have a different action: thus, ferritin has a pro-oxidant capacity, while hemosiderin and lactoferrin are antioxidants.

Bilirubin, resulting from the metabolism of hemoglobin, as transition metal ions, causes alterations in the membrane structure by initiating PUFA peroxidation. Bilirubin crosses the blood-brain barrier, inhibits oxidative phosphorylation and decreases AMPc and GSH concentration. Thus, the encephalopathy caused by intense hemolytic jaundice in neonates is correlated with elevated levels of bilirubin, blood lipid peroxides and GSH decrease.

The same changes were observed in hepatitis of various etiologies (viral, ethanolic) and were correlated with graded morphological changes of the steatosis type, up to the irreversible ones, cirrhosis, caused by exceeding the protective antioxidant systems.

The bilirubin has an antioxidant effect, enhanced by binding to albumin, its plasma transport form. This different behavior of bilirubin depends on the concentration and the environment, like ascorbic acid, which features a pro- and antioxidant character, widely accepted today.

6. Reactive oxygen species – Implications in cardiovascular pathology

Atherosclerosis (ATS) and its notable complication, coronary heart disease, still represent the major cause of premature death worldwide. Several lines of evidence suggest that the major risk factors (hypertension, diabetes mellitus, hyperlipemia, smoking) elicit oxidative stress at the luminal surface of vascular wall that will be further responsible for the oxidative damage of lipoproteins, formation of lipid peroxides, platelet aggregation and activation of macrophages [10]. LDL lipoproteins are the easiest to be oxidized because of their high PUFA content; at variance from native LDL, oxidatively modified LDLs are more avidly taken up by macrophages via the scavenger receptor thus generating the well-known "foam cells" of the atherosclerotic plaques. Experimental studies demonstrated that LDL can be oxidized by all of the major cells of the arterial wall (macrophages, endothelial cells, smooth muscle cells). Besides its rapid uptake by macrophages, oxidized LDL elicit a chemoatractive effect facilitating monocyte adhesion to the endothelium and a toxic affect at the level of endothelial cells by inhibiting the release of nitric oxide. In vivo identification of oxidized LDL in atherosclerotic plaques clearly established in the late 80s the oxidative-modification theory of ATS. Much effort was further directed towards identification of factors that influence the susceptibility of LDL particles to oxidation. Among these, the presence of small dense LDS particles, of preformed lipid peroxides, as well as glycation or binding of LDLs to proteoglycans were proven to facilitate oxidation [12].

Highly reactive aldehydes are one of the major causative factors in oxidative related cardiovascular pathology and ageing. Specific aldehydes (e.g., 4-hydroxynonenal acetaldehyde, acrolein) were reported to be transiently increased in the settings of heart failure and ischemia-reperfusion injury [13] and to interfere with transcriptional regulation of endogenous anti-oxidant networks in mitochondria [1]. Recently, accumulation of reactive aldehydes was studied from the point of view of the subsequent protein carbonylation and its implication in cardiovascular pathophysiology [4].

On the other hand, decreased antioxidant defense further contributes to the oxidative damage. Low concentration of GSH-peroxidase in the vascular wall creates conditions favorable to the actions of hydrogen peroxide and other FR on lipids and lipoproteins [28]. In physiological conditions, nitric oxide acts as an antioxidant, inhibiting LDL peroxidation and their destructive effect on interstitial proteoglycans. With the increased production of FR, NO may become a prooxidant factor, stimulating LDL peroxidation by a mechanism involving myoglobin. Deficiency of other protective factors will favor oxidative injury. Lipid-soluble antioxidants such as tocopherols and ubiquinol are present in the hydrophobic environment of the lipoproteins in order to protect PUFA from FR attack. *In vitro* experimental data showed that: i) exposure of LDL to oxidative stress will trigger lipid peroxidation only after the loss of its above mentioned antioxidants and ii) enrichment of LDL with vitamin E will make LDL oxidation more difficult [6].

Accordingly, the beneficial role of antioxidant supplementation has been extensively investigated in the past decades in a variety of animal models. Most investigators reported beneficial effects, i.e., prevention of atherosclerotic lesions with vitamin E supplementation, yet an early study by Keaney et al. mentioned a deleterious effect of high doses of tocopherol on endothelial-dependent relaxation in cholesterol fed rabbits [11]. Unfortunately, despite the promising observational experimental data, several prospective, double-blind, placebo-controlled trials did not support a causal relationship between vitamin C and E supplementation and a lower risk of coronary heart disease [21]. Similarly, lack of beneficial effect with long term vitamin E supplementation was recently reported in large clinical trial (the Women's Health Study) that addressed the role of antioxidant therapy in the primary prevention of heart failure [2].

These negative results may be related to the fact that antioxidant supplements could abolish the physiological role of ROS as signaling molecules [18], especially when considering that most cardiovascular patients are treated with "pleiotropic" drugs such as statins, angiotensin-converting enzyme inhibitors, angiotensin receptor blockers, that besides their major effects are reported to reduce ROS formation [23]. Indeed, a large body of evidence demonstrated unequivocally that reduced amounts of reactive oxgen species, most probably of mitochondrial origin [17] but not exclusively, are essential in regulating cardiovascular homeostasis [19] as well as the powerful mechanisms of endogenous cardioprotection at postischemic reperfusion, namely pre- and postconditioning [20].

In conclusion, increasing the level of endogenous antioxidants, as recently suggested via the supplimentation of weak "pro-oxidants" [8], and not chronic supplementation with large dose of exogenous antioxidants could become in the future a more appropriate approach to treat diseases that share oxidative stress as a common denominator.

Author details

Claudia Borza, Danina Muntean, Cristina Dehelean,
Germaine Săvoiu, Corina Şerban, Georgeta Simu, Mihaiela Andoni,
Marius Butur and Simona Drăgan
University of Medicine and Pharmacy "Victor Babeş" Timişoara, Romania

7. References

[1] Anderson EJ, Katunga LA, Willis MS. Mitochondria as a source and target of lipid peroxidation products in healthy and diseased heart. *Clin Exp Pharmacol Physiol* 2012, 39(2): 179-93.

[2] Chae CU, Albert CM, Moorthy MV, Lee IM, Buring JE – Vitamin E supplementation and the risk of heart failure in women *Circ Heart Fail* 2012, 5(2): 176-82.

[3] Cheesman KH, Slater TF - An introduction to free radical biochemistry, in *Free radicals Medicine: British Medical Bulletin*, 1993, vol 49, 3, 481-494.

[4] Fritz KS, Petersen DR. Exploring the biology of lipid peroxidation-derived protein carbonylation. *Chem Res Toxicol* 2011, 24(9): 1411-1419.

[5] Fulbert J C, Cals M-J - Les radicaux libres en biologie clinique: origine, role pathogene et moyens de defense, in *Pathologie Biology*, 1992, vol 40, 1, 66-77.

[6] Gutteridge JM, Halliwell B. Antioxidants: Molecules, medicines, and myths. *Biochem Biophys Res Commun* 2010, 393(4): 561-564.

[7] Halliwel B, Gutteridge C, Cross C - Free radicals, antioxidants and human disease. Where are you now?, *J Lab Clin Med*, 1993, 119 (6), 606-610.

[8] Halliwell B. Free radicals and antioxidants: updating a personal view. *Nutr Rev* 2012, 70(5): 257-265.

[9] Holley A, Cheesman K H - Measuring free radical reactions in vivo, in *Free radicals in Medicine: British Medical Bulletin*, 1993, vol 49, 3, 494-506.

[10] 10.Kawada T - Oxidative stress markers and cardiovascular disease: advantage of using these factors in combination with lifestyle factors for cardiovascular risk assessment. *Int J Cardiol* 2012, 157(1): 119-120.

[11] Keaney JF Jr, Gaziano JM, Xu A, Frei B, Curran-Celentano J, Shwaery GT, Loscalzo J, Vita JA – Low dose alpha-tocopherol improves and high-dose alpha tocopherol worsens endothelial dependent vasodilator function in cholesterol-fed rabbits. *Journal of Clinical Investigation*, 1994, 93(2):844-851.

[12] Lee R, Margaritis M, Channon KM, Antoniades C - Evaluating oxidative stress in human cardiovascular disease: methodological aspects and considerations. *Curr Med Chem* 2012, 19(16): 2504-20.

[13] Marczin N, El-Habashi N, Hoare GS, Bundy RE, Yacoub M - Antioxidants in myocardial ischemia-reperfusion injury: therapeutic potential and basic mechanisms. *Arch Biochem Biophys* 2003, 420 (2), 222-236.

[14] Mederle C, Raica M, Schneider F, Tănăsie G - The constrictor effect of acethylcoline by enhanced hydrogen peroxide in ovalbumin - sensitised rat tracheal muscle. *The Romanian - Hungarian Physiology Joint - Meeting, 1996, Szeged - Timişoara, Physiology,* 6, 2 (10).

[15] Mederle C, Schneider F, Mătieş R - Oxidative stress evaluation in experimental and clinical asthma. *Inter Rew Allerg Clinic Immunol,* III, 1997,1: 30-31.

[16] Mederle C, Schneider F, Raica M - Free radicals and mast cell reaction in tracheal muscle of rats sensitised with ovalbumin. *Allergy* (suppl), 1996, 31, 51: 188-189.

[17] Muntean DM, Mirica SN, Ordodi V, Răducan AM, Duicu OM, Gheorghiu G, Henţia C, Fira-Mlădinescu O, Săndesc D – Mitocondria: novel therapeutic target in cardioprotection. *Rev Rom Med Vet* 2008, 1(4): 406-412.

[18] Murphy MP, Holmgren A, Larsson NG, Halliwell B, Chang CJ, et al. Unraveling the biological roles of reactive oxygen species. *Cell Metab* 2011, 13(4): 361-6

[19] Otani H. Reactive oxygen species as mediators of signal transduction in ischemic preconditioning. *Antiox Redox Signal* 2004, 6(2): 449-469.

[20] Penna C, Mancardi D, Rastaldo R, Pagliaro P - Cardioprotection: a radical view. Free radicals in pre- and postconditioning. *Biochim Biophys Acta,* 2009, 1787(7): 781-93.

[21] Riccioni G, Frigiola A, Pasquale S, Massimo deG, D'Orazio N - Vitamin C and E consumption and coronary heart disease in men. *Front Biosci* 2012, 1(4): 373-380.

[22] Schneider F, Mederle C, Mătieş R, Săvoiu G, Goţia S, Mihalaş G I, Buftea A - Drug influence of morphological and functional response in contention stress by rats. *Fiziologia - Physiology,* 1992, 2, 2:27.

[23] Selvaraju V, Joshi M, Suresh S, Sanchez JA, Maulik N, Maulik G. Diabetes, oxidative stress, molecular mechanism, and cardiovascular disease - an overview. *Toxicology Mech Met* 2012, 22(5): 330-5.

[24] Sies H - Oxidative stress: From Basic Research to Clinical Application: *The American Journal of Medicine: Proceedings of a Symposium: Oxidants and Antioxidants: Pathophysiological Determinants and Therapeutic Agents:* vol 91 (3C), 1991, (3C-31S).

[25] Tinkel J, Hassanain H, Khouri SJ - Cardiovascular antioxidant therapy: a review of supplements, pharmacotherapies, and mechanisms. *Cardiol Rev* 2012, 20(2): 77-83.

[26] Valko M, Leibfritz D, Moncol J, Cronin M, Mazur M, Telser J. Free radicals and antioxidants in normal physiological functions and human disease. *Int J Biochem & Cell Biol* 2007, 39, 44-84.

[27] Verceloti M G, Severson P S, Duane P, Moldow F - Hydrogen peroxide alters signal transduction in human endothelial cells. *J Lab Clin Med*, 1991, 117,1: 15-23.

[28] Yung LM, Leung FP, Yao X, Chen ZY, Huang Y. Reactive oxygen species in vascular wall. *Cardiovasc Hematol Disord Drug Tagets*. 2006, 6(1): 1-19.

Chapter 3

The Role of Liver X Receptor in Hepatic *de novo* Lipogenesis and Cross-Talk with Insulin and Glucose Signaling

Line M. Grønning-Wang, Christian Bindesbøll and Hilde I. Nebb

Additional information is available at the end of the chapter

1. Introduction

Regulation of nutrient balance by the liver is important to ensure whole body metabolic control. Hepatic expression of genes involved in lipid and glucose metabolism is tightly regulated in response to nutritional cues, such as glucose and insulin. In response to dietary carbohydrates, the liver converts excess glucose into fat for storage through *de novo* lipogenesis. The liver X receptors (LXRα and LXRβ) are important transcriptional regulators of this process. LXRs are classically known as oxysterol sensing nuclear receptors that heterodimerize with the retinoic X receptor (RXR) family and activate transcription of nutrient sensing transcription factors such as sterol regulatory element-binding protein 1c (SREBP1c) (Repa et al., 2000; Yoshikawa et al., 2002; Liang et al., 2002) and carbohydrate response element-binding protein (ChREBP) (Cha & Repa, 2007). LXR also induces the transcription of the lipogenic enzyme genes fatty acid synthase (FAS), stearoyl-Coenzyme A desaturase (SCD1) and Acetyl CoA carboxylase (ACC), alone or in concert with SREBP1c and/or ChREBP (Chu et al., 2006; Joseph et al., 2002; Talukdar & Hillgartner, 2006). LXR activate transcription of hepatic lipogenic genes in response to feeding, which is believed to be mediated by insulin (Tobin et al., 2002). The mechanisms by which insulin activates LXR-mediated gene expression is not clearly understood, but may involve production of endogenous ligand for LXRs and/or act by signal transduction mechanisms downstream of the insulin receptor (IR). Both glucose and insulin regulate *de novo* lipogenesis, however, some lipogenic genes can be regulated by glucose without the need of insulin which has been shown for SREBP1c (Hasty et al., 2000; Matsuzaka et al., 2004). A well known glucose-mediator in liver is ChREBP, an important regulator of *de novo* lipogenesis in response to glucose (Yamashita et al., 2001). ChREBP is activated by glucose via hexose- and pentose-phosphate-dependent mechanisms involving dephosphorylation of ChREBP and translocation to the nucleus (Havula & Hietakangas, 2012). Interestingly, both LXR and ChREBP were recently shown to be post-translationally modified by O-linked β-N-acetylglucosamine (O-GlcNAc) in response to glucose potentiating their lipogenic capacity (Anthonisen et al., 2010; Guinez et al., 2011). Glucose flux through the hexosamine signaling pathway generates UDP-N-acetyl-glucosamine (UDP-GlcNAc), a substrate for O-GlcNAc modification of nucleocytoplasmic proteins by the enzyme O-GlcNAc transferase (OGT). We have shown that O-GlcNAcylation of LXR is increased in mouse livers in response to feeding and in livers from hyperglycemic diabetic mice potentiating SREBP1c expression (Anthonisen et al., 2010). Furthermore, preliminary studies in our laboratory indicate that LXR potentiate ChREBP activity under hyperglycemic conditions establishing a link between glucose metabolism, LXR and ChREBP. These observations suggest that LXR, SREBP1c and ChREBP contribute to converting carbohydrates into fat in a cooperative manner in response to high circulating glucose levels and that O-GlcNAc signaling plays a role in this process. As O-GlcNAc cycling appear to be essential for proper insulin signaling and the sensitivity of OGT to glucose increases with decreasing insulin signaling (Mondoux et al., 2011; Hanover et al., 2010) the relative roles of LXR, SREBP1c and ChREBP in regulating *de novo* lipogenesis in response to feeding and modification by O-GlcNAc signaling under insulin sensitive and insulin resistant conditions will be discussed.

2. Liver X Receptors (LXR)

2.1. LXR structure and function

LXRα (NR1H3) and LXRβ (NR1H2) are ligand-activated transcription factors belonging to the nuclear receptor (NR) superfamily (Lehmann JM (Lehmann et al., 1997; Willy et al., 1995; Janowski et al., 1996). LXRα is primarily expressed in metabolically active tissues, such as liver, intestine, adipose tissue, kidney and macrophages, whereas LXRβ is ubiquitously expressed (Apfel et al., 1994; Teboul et al., 1995; Teboul et al., 1995). LXRs are intracellular sensors of cholesterol and oxidized cholesterol derivatives (oxysterols) have been identified as their endogenous ligands (Janowski et al., 1996; Lehmann et al., 1997). The two isotypes originates from two different genes on separate chromosomes, but share the same modular structure, which is characteristic of most NRs (Fig. 1).

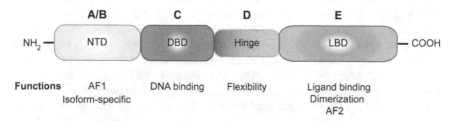

Figure 1. Structure of the LXRs

The DNA-binding domain (DBD) and the ligand binding domain (LBD) are highly structured domains. LXRα and LXRβ share 78 % amino acid sequence identity in these regions, while the N-terminal domain (NTD) and the hinge domain are far more disordered and less conserved. DNA binding requires dimerization with RXR. Transactivation by the LXRs is mediated through the ligand independent activation function (AF1) in NTD and the ligand dependent activation function 2 (AF2) in the LBD. Binding of a ligand to the hydrophobic ligand binding pocket leads to a conformational change that releases corepressors (CR) and exposes binding sites for coactivators (CA), recruiting the general transcription machinery and RNA polymerase II (RNA Pol II) (Fig. 2). This leads to changes in LXR dependent gene expression. The interactions with coregulators can also occur independently of ligand to AF1, however this is far less characterized. Upon activation, LXRs regulate a number of genes involved in lipid, cholesterol and glucose metabolism by binding to LXR response elements (LXREs) in their promoter region. These consist of a direct repeat of the nucleotide hexamer AGGTCA spaced by four nucleotides. Insights into LXR function in metabolism was provided by the generation of LXR mutant mice. These mice accumulate hepatic cholesterol, ultimately causing liver dysfunction (Peet et al., 1998; Ulven et al., 2005). It was found that LXRα controls cholesterol metabolism by conversion of cholesterol to bile acid by induction of the cholesterol 7 alpha-hydroxylase (Cyp7A1) gene, biliary cholesterol excretion and cholesterol efflux via induction of ABCG5/8 and ABCA1/ABCG1, respectively (Lehmann et al., 1997; Chiang et al., 2001; Yu et al., 2003; Repa et al., 2002; Graf et al., 2002; Costet et al., 2000; Sabol et al., 2005; Venkateswaran et al., 2000; Venkateswaran et al., 2000). LXRs are strongly implicated in the development of metabolic disorders and associated pathologies, notably, hyperlipidemia and atherosclerosis (Peet et al., 1998; Calkin & Tontonoz, 2010). Thus, LXRs are key players in maintaining metabolic homeostasis in health and disease by regulating inflammation and lipid/carbohydrate metabolism.

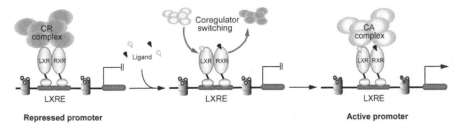

Figure 2. Activation of LXR by coregulator switching

2.2. Modulation of LXR activity by coregulators and PTMs

The transcriptional activity of LXRs is highly dependent on the presence of coregulators which has been linked to several metabolic processes (Jakobsson et al., 2009; Kim et al., 2003; Huuskonen et al., 2004; Kim et al., 2008; Oberkofler et al., 2003). Coregulators constitutes large multisubunit protein complexes containing chromatin-remodelling and/or –modifying enzymes with intrinsic histone acetylase (HAT)/ deacetylase (HDAC) and histone methylase (HMT)/demethylase (HDM) activities, depending on whether they act as activators or repressors, respectively (Kato et al., 2011). It has been assumed that that the unliganded LXRs are localized in the nucleus and interact with CRs, including nuclear receptor corepressor/silencing mediator of retinoic acid and thyroid receptor (NcoR/SMRT) (Wagner et al., 2003). However, recent chromatin immunoprecipitation (ChIP) studies, including ChIP-sequencing (ChIP-Seq), have challenged this classical model. These studies put forward a more complex view, that ligands, pioneer factors, coregulators and posttranslational modifications (PTMs) play different roles in determining the LXR binding sites and actions *in vivo* (Boergesen et al., 2012; Heinz et al., 2010; Pehkonen et al., 2012). Furthermore, some coregulators have been shown to act as dual function activators/repressors, such as the coregulator protein receptor interacting protein 140 (RIP140). RIP140 has been shown to serve as a CA for LXR in lipogenesis but as a CR in gluconeogenesis independent of ligand activation (Herzog et al., 2007). General mechanisms of coregulator actions are assumed to be conserved between LXRs, but based on the low amino acid sequence identity in the NTD (32%) and the hinge domain (25%) it is possible that they contain novel isotype specific interaction surfaces. Also, the specific coregulator requirement to lipogenic LXR target genes in response to different feeding regiments under normal and diabetic conditions remain largely unexplored. In addition to ligand binding, LXRs can be posttranslationally modified by phosphorylation, acetylation, and sumoylation, affecting their target gene specificity, stability, and transactivating and transrepressional activity, respectively (Li et al., 2007; Ghisletti et al., 2007; Chen et al., 2006; Yamamoto et al., 2007). We have recently shown that LXR can be modified by O-GlcNAcylation in response to glucose (see section 4.3), increasing its transactivation of the SREBP1c promoter (Anthonisen et al., 2010). PTMs may alter the structural conformation of LXR thereby modifying the affinity of coregulators that determines whether a target gene is induced or suppressed. Modulation by PTMs can occur both in the absence and presence of natural ligand tuning LXR activities in a cell- and gene-specific manner (Rosenfeld et al., 2006) depending on the nutritional stimuli.

3. LXR in hepatic *de novo* lipogenesis

3.1. LXR lipogenic target genes

In addition to being central regulators of cholesterol metabolism, the LXRs are involved in induction of fatty acid and triglyceride (TG) biosynthesis in response to feeding. *De novo* lipogenesis ensures that excess acetyl-CoA, which is an intermediate product of glucose metabolism, is converted into fats and subsequent TGs. LXRs are involved in hepatic lipogenesis through direct regulation of SREBP1c and ChREBP expression (Repa et al., 2000; Cha & Repa, 2007; Shimano, 2001). SREBP1c is a well described transcriptional regulator of hepatic lipogenesis (Shimano, 2001), and together with LXR and glucose-regulated ChREBP (see section 4.1), it controls expression of essential enzymes in lipogenesis, lipid storage and secretion (Fig. 3). SREBP1c deficiency does not fully abolish the expression of genes involved

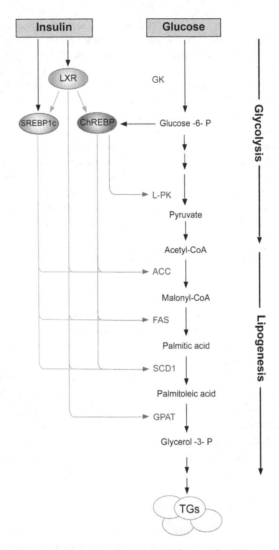

Figure 3. Regulation of hepatic lipogenesis by LXR, SREBP1c and ChREBP.

in hepatic lipogenesis. SREBP1c null mice treated with an LXR agonist results in induction
of a subset of lipogenic genes and a modest increase in fatty acid synthesis (Liang et al.,
2002), which implies that LXR can act independently of SREBP1c. In particular, the SCD1
gene is directly regulated by LXRα in response to synthetic ligands, also in the absence of
SREBP1c (Chu et al., 2006). SCD1 is central in desaturation of saturated fatty acyl-CoAs
important for formation of cholesterol esters (CEs) and TGs. Thus, specific LXR-mediated
regulation of SCD1 can be explained by the essential role of LXR in limiting toxic free

cholesterol in response to diets rich in cholesterol and saturated fat. The expression of LXRα in liver is rapidly upregulated by insulin *in vivo*, increasing mRNA expression of SREBP1c, malic enzyme (ME), ACC and FAS. Furthermore, expression of these lipogenic genes was abolished in insulin-injected LXRα/β double knock out mice (Tobin et al., 2002), indicating an essential role for LXR in insulin-mediated regulation of hepatic lipogenesis. The mechanisms by which insulin activate LXR-mediated gene expression is not clearly understood, but may involve production of endogenous ligand for LXRα/β (Chen et al., 2004) and/or by signal transduction mechanisms downstream of the IR affecting CA recruitment to LXRs and/or PTMs of LXRs. This will be discussed in more detail below. Of note, PKA-induced phosphorylation of LXRα has been shown to inhibit the expression of SREBP1c in liver from mice via reduced DNA binding and CA recruitment (Yamamoto et al., 2007). Since glucagon/cAMP/PKA signaling may, at least in part, explain down-regulation of SREBP1c expression in response to fasting, it is likely that PKA-mediated phosphorylation of LXR contributes to the fasting signal on SREBP1c.

3.2. Putative mechanisms regulating LXR-mediated *de novo* lipogenesis in response to insulin

Insulin is the most important anabolic hormone in the body, regulating many processes important for cellular growth and energy storage such as glucose uptake and metabolism, glycogen and lipid synthesis, gene transcription and translation. A classic action of insulin is to mediate a metabolic switch from fatty acid oxidation to synthesis and suppress hepatic glycogenolysis and gluconeogenesis in response to carbohydrate excess, a process that is largely regulated at the transcriptional level. In this way, hepatic insulin signaling maintains whole body energy homeostasis. In the insulin-resistant state, only the ability of insulin to suppress hepatic gluconeogenesis is lost, while its ability to activate lipogenesis is retained (Shimomura et al., 2000; Matsumoto et al., 2006; Brown & Goldstein, 2008). This bifurcated insulin resistance can be explained by failure of insulin to inhibit the gluconeogenic transcription factor Forkhead box protein O1 (FoxO1), but maintaining signaling to lipogenic transcriptional regulators including LXR and SREBP1c.

3.2.1. The insulin signaling cascade

The insulin signaling cascade is initiated by the binding of insulin to the extracellular β-subunits of the dimerized IR followed by autophosphorylation on several intracellular tyrosine residues on the IR. Insulin receptor substrate (IRS) is an essential protein docking onto the phosphorylated IR which in turn is phosphorylated itself on multiple tyrosine residues. This creates docking sites for src homology 2 (SH2) domain containing proteins. The best studied SH2 protein that binds to tyrosine phosphorylated IRS proteins is the regulatory subunit of the phosphoinositide 3-kinase (PI3K). PI3K catalyzes the formation of the lipid second messenger phosphatidylinositol (3,4,5) trisphosphate (PIP3), which is necessary to recruit downstream kinases. PIP3 generates a binding site for proteins containing Pleckstrin homology (PH) domains, such as 3′-phosphoinositide-dependent protein kinase (PDK1), the serine/threonine kinase Akt/protein kinase B and possibly also mammalian target of rapamycin complex 2 (mTORC2). PDK and mTORC2 are both necessary for full activation of Akt downstream of the insulin receptor via PDK1-mediated phosphorylation of Akt on threonine 308 and mTORC2-mediated phosphorylation on serine 472 (Saltiel & Kahn, 2001; White, 2003; Jacinto et al., 2006). All these events occur transiently in specific cholesterol rich plasma membrane microdomains called caveolae, generating a specific signaling unit for proper downstream insulin signaling where Akt plays a central role.

3.2.2. Regulation by mTOR

One of the targets of Akt is mTORC1 (Zoncu et al., 2011). Recent evidence suggests that mTORC1 is involved in LXR-mediated lipogenic gene transcription including induction of SREBP1c, FAS and ACC in liver from mice subjected to a high fat diet (Hwahng et al., 2009). The authors show that the mechanism by which mTORC1 activates LXR is via p70 S6 kinase (S6K)-mediated phosphorylation of LXR. Conversely, in the fasted state, LXR was shown to be inhibited by AMPK-mediated phosphorylation. In agreement with these observations, Li et al (Li et al., 2010) showed that insulin-activated hepatic transcription of SREBP1c, FAS and SCD1 is mediated by mTORC1, however independent of S6K. As both LXR and SREBP1c induce lipogenic promoters in response to insulin, this might suggest that activation of LXR in response to insulin/nutrients is mediated, at least in part, by mTORC1 and S6K, whereas insulin-signaling to SREBP1c requires mTORC2 independently of S6K, possibly via Akt-mediated inhibition of glycogen synthase kinase-3 (GSK3) (Hagiwara et al., 2012). In this way, GSK3-mediated phosphorylation and degradation of SREBP1c is prevented by insulin signaling to mTORC2 and Akt. Of note, insulin has primarily been shown to act on the SREBP1c promoter by activating LXRs and not SREBP1c (Chen et al., 2004) and the effect of insulin on SREBP1c is mainly at the posttranslational level. In a recent publication, mTORC1 was shown to phosphorylate a phosphatidic acid phosphatase, Lipin 1, preventing its nuclear entry and subsequent inhibition of SREBP1c-mediated activation of the FAS promoter (Peterson et al., 2011). Furthermore, Yecies JL et al (Yecies et al., 2011) showed that Akt2 independently of mTORC1 downregulate the mRNA expression of insulin induced gene 2 (Insig2a), an inhibitor of SREBP1c. This finding has been debated by Wan M et al (Wan et al., 2011), who could not observe any downregulation of Insig2a by Akt2. They postulate that Akt2 acts independently of mTORC1 and SREBP1c, possibly via posttranslational mechanisms, and that nutrients have a direct role in the liver to promote lipogenesis by a process dependent on both mTORC1 and other insulin-dependent signaling pathways. In light of the above mentioned studies, both mTORC1 and mTORC2 (Soukas et al., 2009; Guertin et al., 2006; Lamming et al., 2012; Hagiwara et al., 2012) appear to play important roles in lipid synthesis and storage in hepatocytes. Further studies will reveal the relative roles of Akt1, Akt2, mTORC1/C2 and S6kinase on activation of LXR and SREBP1c in this regulation under insulin sensitive and insulin resistant conditions and cross-talk with glucose metabolism and signaling (Fig.4).

3.2.3. Regulation by FoxO1

Another mechanism by which insulin may promote LXR-mediated SREBP1c transcription is through the transcription factor FoxO1. FoxO1, generally known as an activator of gluconeogenic genes during fasting, can repress the transactivating ability of LXR and cooperating transcription factors SREBP1c and Specificity protein 1 (Sp1) to activate SREBP1c transcription during fasting (Liu et al., 2010; Deng et al., 2012). FoxO1 does not seem to bind directly to the SREBP1c promoter, but appears to act as a repressor through protein-protein interactions, possibly by recruiting CR proteins (Deng et al., 2012). Upon feeding, FoxO1 is inhibited by insulin via PI3-kinase activation and phosphorylation by Akt, which excludes phosphorylated FoxO1 from the nucleus via association with the 14-3-3 protein (reviewed in (Tzivion et al., 2011)). In this way, at least under insulin sensitive conditions, inhibition mediated by FoxO1 and associating CRs is relieved, enabling LXR, Sp1 and SREBP1c to activate the SREBP1c promoter in a cooperative fashion. Of note, an important role for the E-box transcription factor Upstream Stimulatory Factor (USF) in mediating insulin activation of the SREBP1c promoter has also been reported (Wong & Sul, 2010). The relative roles of LXR, SREBP1c and cooperating transcription factors in regulation of the SREBP1c promoter after high-carbohydrate feeding under normal and insulin resistant conditions and the role of FoxO1 in this process in insulin resistance is currently not known. Recently, the role of Akt as a central regulator of both gluconeogenesis, through inhibition of FoxO1, and lipogenesis, through activation of mTORC1/2 in hepatic insulin signaling, was debated as the insulin resistant phenotype of mice lacking hepatic Akt1/2 were normalized in mice with concomitant liver-specific deletion of FoxO1 (Lu et al., 2012). This work suggests that a major role for Akt as a metabolic regulator in response to insulin is largely to restrain FoxO1 activity, at least for suppression of liver glucose output.

3.2.4. Regulation by insulin-mediated oxysterol production

Considering the bifurcated nature of insulin resistance and the postulated central role of Akt in this process, a very recent work by Wu and Williams (Wu & Williams, 2012), put forward an interesting theory. They suggest that disturbance of a single molecule, NAD(P)H oxidase 4 (NOX4), is sufficient to induce the key harmful features of insulin resistance. NOX4 is activated upon IR activation, generating a transient burst of superoxide (O_2^-) and its byproduct H_2O_2. This enhances signal transduction by disabling enzymes in the protein-tyrosine phosphatase gene family. In this way, essential inhibiting enzymes in the insulin signaling cascade is blocked, notably the PI3K inhibitor PTEN and protein-tyrosine phosphatase-1B (PTP1B) (Wu & Williams, 2012). Intriguingly, NOX4 may also be the link between insulin signaling and production of oxysterol ligand for LXR, as NOX4 through its superoxide producing activity may mediate the production of oxygenated cholesterol. The evidence for this is that pharmacological inhibition of NOX4 blocked insulin-induction of SREBP1c mRNA in rat primary hepatocytes, even though phosphorylations upstream and downstream of mTORC1 remained responsive (Wu & Williams, 2012). Furthermore, NOX4 is transiently localized to caveolae (Han et al., 2012), possibly via recruitment to the IR, placing the enzyme in close proximity to cholesterol-rich areas of the plasma membrane. A complete summary of putative mechanisms of insulin-mediated signaling to LXR, SREBP1c and lipogenesis is depicted in Fig. 4.

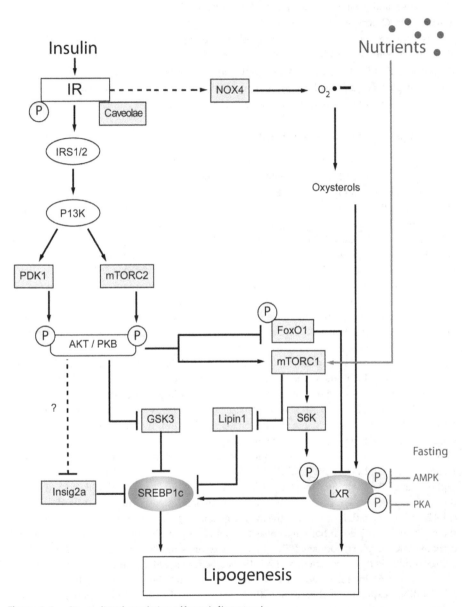

Figure 4. Insulin-mediated regulation of hepatic lipogenesis

4. Lipogenic gene expression in response to glucose metabolism

Hepatic glucose metabolism activates the transcription of various genes encoding enzymes of glycolysis and lipogenesis independently of insulin. However, the initial modification of glucose into Glucose-6-phosphate (G6P) by the enzyme Glucokinase (GK; Hexokinase 4) required for transcriptional regulation by glucose is highly dependent on insulin (Bosco et al., 2000), possibly via SREBP1c (Foretz et al., 1999; Kim et al., 2004) in concert with LXR and Peroxisome Proliferator-Activated Receptor gamma (PPARγ) (Kim et al., 2009). Thus the actions of glucose and insulin may be considered interdependent and that regulation of gene expression in response to glucose seems to require active LXR, SREBP1c and/or PPARγ.

4.1. Glucose regulation via ChREBP

A majority of hepatic glucose-responsive genes is thought to be regulated by the transcription factor ChREBP (Yamashita et al., 2001; Ishii et al., 2004). ChREBP mediates transcriptional regulation of glycolytic and lipogenic enzymes and is particularly important for the induction of liver-pyruvate kinase (L-PK), one of the rate limiting enzymes of glycolysis, which is exclusively dependent on glucose (Matsuda et al., 1990; Dentin et al., 2004). Furthermore, ChREBP is involved in regulating ACC and FAS in concert with LXR and SREBP1c in response to glucose and insulin, respectively, suggesting its involvement of the conversion of carbohydrates into fat (Joseph et al., 2002; Talukdar & Hillgartner, 2006). Moreover, stimulation by a synthetic LXR ligand, induces hepatic expression and activity of ChREBP (Cha & Repa, 2007). However, ChREBP is apparently not dependent on LXR for its hepatic expression and activity in mice fed a high carbohydrate/high fat diet (Denechaud et al., 2008), suggesting that ChREBP activity is reinforced by upstream LXR under certain nutritional conditions. At low glucose concentrations, the ChREBP protein is retained as an inactive phosphoprotein in the cytoplasm (reviewed in (Havula & Hietakangas, 2012)). The mechanisms by which glucose activate ChREBP is not clear, but involves induction of the ChREBP mRNA, dephosphorylation of the protein, shuttling to the nucleus and binding to the ChREBP response element at the promoter of its target genes (Uyeda & Repa, 2006). Early studies pointed to xylose 5-phosphate (Xu5P), an intermediate of the pentose phosphate pathway (PPP), as an activating signal through its ability to activate protein phosphatase 2A (PP2A) and subsequent dephosphorylation of ChREBP (Havula & Hietakangas, 2012). Recently, ChREBP was shown to be activated by fructose 2,6-biphosphate (F2,6BP) in hepatocytes (Arden et al., 2012). The level of F2,6BP is regulated by the bifunctional enzyme 6-phosphofructokinase-2-kinase/fructose-2,6-biphosphatase (PFK2/FBP2). Thus, PFK2 catalyzes the synthesis and degradation of F2,6BP and as a result, the enzyme is involved in both glycolysis and gluconeogenesis. In the fed state, insulin and carbohydrates dephosphorylate PFK2 in the liver making the enzyme kinase dominant. Subsequently, F6P is converted to F2,6BP that activates PFK1, which in turn stimulates glycolysis (Fig. 6). Interestingly, LXRα was recently shown to be a central regulator of hepatic PFK2 mRNA expression (Zhao et al., 2012). Activation of ChREBP in response to glucose appears to depend on multiple glucose metabolites, including G6P, X5P and F2,6BP. As LXRα is involved in regulation GK- and PFK2-expression in response to insulin, this may suggest that ChREBP is dependent on insulin signaling via LXR for proper substrate availability.

4.2. Glucose metabolism via the hexosamine biosynthetic pathway and O-GlcNAc signaling

Glucose metabolism from F6P can follow the alternative hexosamine biosynthetic pathway (HBP) where the enzyme glutamine fructose-6-phosphate amidotransferase (GFAT) controls the first and rate limiting step (Fig. 5).

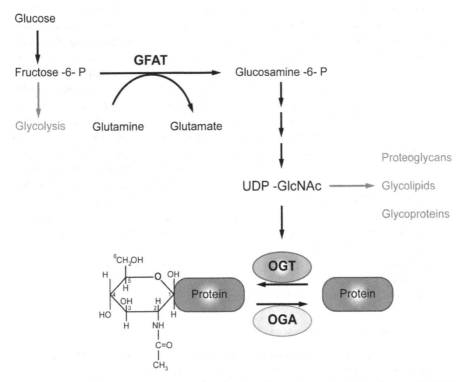

Figure 5. Nutrient flux and O-GlcNAc modification of nucleocytoplasmatic proteins through the HBP

The end product of this pathway is Uridine diphosphate N-acetylglucosamine (UDP-GlcNAc), an essential building block for N-and O-linked glycosylation of proteins and lipids. Cytoplasmic and nuclear proteins can be dynamically modified by O-linked β-N-acetylglucosamine (O-GlcNAc) on serine and threonine residues by the enzyme O-GlcNAc transferase (OGT) using UDP-GlcNAc as substrate. OGT is an essential enzyme as targeted deletion of this gene is lethal (Shafi et al., 2000). The enzyme O-GlcNAc transferase (OGA) hydrolyses the sugar analogous to protein dephosphorylation of phosphorylated proteins by phosphatases (Hart et al., 2007; Love, 2005). Because O-GlcNAc levels on proteins appear to be sensitive to increasing flux through this pathway in response to nutrient excess, OGT can be considered as a general sensor of glucose availability that modifies proteins according to changes in UDP-GlcNAc levels. There is no identified consensus sequence for GlcNAcylation, and unlike the multiple genes encoding kinases, there is only a single X-linked gene encoding the catalytic subunit of OGT in mammals (Shafi et al., 2000). For this reason, it has been hypothesized that OGT is the catalytic subunit in large transient enzyme complexes where interacting proteins are able to target OGT to its many substrates. Many transcription factors are modified by O-GlcNAc in the liver (Dentin et al., 2008; Housley et al., 2008; Kuo et al., 2008; Ozcan et al., 2010). Interestingly, FoxO1 has been shown to be a target for O-GlcNAcylation in hepatocytes in response to hyperglycemia in the insulin resistant state, resulting in elevated transactivating capacity for FoxO1 against its gluconeogenic targets phosphoenolpyruvate carboxykinase (PEPCK) and glucose-6-phosphatase (G6Pase) reinforcing hepatic glucose production (Housley et al., 2008; Kuo et al., 2008). Moreover, this activation was later shown to be dependent on targeting of OGT to FoxO1 via interaction with the coactivator PGC1α, which itself was shown to be modified by O-GlcNAc upon interaction with OGT (Housley et al., 2009). As PGC1α have been shown to significantly amplify LXRα-mediated activation of the SREBP1c promoter (Oberkofler et al., 2003; Kim et al., 2008), a possible recruitment of an OGT/PGC1α-complex to LXR on lipogenic target genes under insulin resistant conditions remains to be explored. Recently, ChREBP was also shown to be a target for O-GlcNAcylation in response to hyperglycemia (Guinez et al., 2011). Adenoviral overexpression of OGT in liver increased ChREBP O-GlcNAc modification, protein stability and transactivating activity of L-PK, as well as potentiating expression of ACC, FAS and SCD1 mRNA expression in response to refeeding (Guinez et al., 2011). In contrast, hepatic overexpression of OGA reduced lipogenic protein content (ACC and FAS) and hepatic steatosis (excessive accumulation of TGs and CEs) in db/db mice, suggesting that enhanced OGT signaling to ChREBP and cooperating transcription factors/coregulators contributes to hepatic steatosis under insulin resistant conditions.

4.3. O-GlcNAc signaling activates LXR and hepatic lipogenesis

In 2007, glucose was reported as an endogenous ligand for LXR (Mitro et al., 2007). This has, however, been debated considering the hydrophobic nature of the ligand binding pocket (Lazar & Willson, 2007). Instead, we asked the question whether glucose exert its effect via hexosamine signaling and posttranslational O-GlcNAc modification of LXR. In a recent publication, we show that LXR is O-GlcNAc modified in response to high glucose (25 mM) in absence of insulin (cells cultured in 2 % serum, approximately 1-2 pmol/l insulin) and synthetic LXR-ligand in Huh7 cells, a human hepatoma cell line (Anthonisen et al., 2010). By pharmacological inhibition we demonstrated that hexosamine signaling and O-GlcNAc cycling mediates LXR dependent activation of the SREBP1c promoter in response to glucose. Furthermore, we observed increased O-GlcNAc modification of LXR in livers from refed mice and streptozotosin (STZ) treated diabetic mice corresponding with increased SREBP1c mRNA expression. Moreover, general protein O-GlcNAcylation was increased in STZ-treated hyperglycemic mice compared to control mice. Our results suggest that LXR is regulated by O-GlcNAc modification, thereby increasing its lipogenic potential. Whether O-GlcNAc-LXR is able to transactivate other lipogenic genes in addition to SREBP1c, is currently under investigation in our laboratory. Our preliminary studies point to a role for O-GlcNAc-LXR in upregulating ChREBP, FAS, ACC and SCD1 expression (Bindesbøll et al, unpublished). Furthermore, preliminary reChIP experiments in our laboratory (LXR ChIP followed by O-GlcNAc ChIP), show a strong induction of O-GlcNAc-associated LXR binding to LXRE on the promoters of SREBP1c, ChREBP, FAS and SCD1 in response to feeding both in control mice and STZ treated mice. Our study is supported by the observation that the SREBP1c promoter activity and protein levels of SREBP1c are increased in response to elevated glucose concentration in the mouse hepatocyte cell line H2-35 (Hasty et al., 2000). Furthermore, treatment with azaserine, an inhibitor of GFAT, completely suppressed expression of both cytoplasmic and nuclear SREBP1c protein, suggesting that hexosamine-dependent O-GlcNAc signaling indeed is involved in glucose-induced SREBP1c mRNA expression, possibly via activation of LXR and/or cooperating transcription factors/CAs.

In our *in vitro* studies, we observed only modest LXR/RXR transactivation of the SREBP1c promoter in high glucose/low insulin-treated cells. This might be explained by constitutive phosphorylation competing for the same site(s) as GlcNAc on LXR and/or inhibitory phosphorylation occurring on adjacent GlcNAc sites. Housley et al. (Housley et al., 2008) reported elevated O-GlcNAc on FoxO1 by high glucose and a subsequent reduction by insulin. They further showed that O-GlcNAc modification increased substantially on the insulin-insensitive mutant FoxO1 lacking three AKT phosphorylation sites (T24A, S256A, S319A), resulting in increased FoxO1-dependent luciferase reporter activity. These observations imply overlapping and/or adjacent phosphorylation and GlcNAc sites on FoxO1. Indeed, the authors also identified several O-GlcNAc sites on FoxO1, one of which is adjacent to an Akt phosphorylation site (Thr[317]). In the case of LXR, which is activated by insulin, apparently in part via S6K-mediated phosphorylation (Hwahng et al., 2009), GlcNAcylation and phosphorylation might act synergistically on LXR in response to glucose and insulin. In fact, extensive cross-talk between O-GlcNAcylation and phosphorylation appear to contribute to the pathology of various diseases (Hart et al., 2011). In addition, GlcNAc and inhibiting phosphate (in response to fasting via PKA and/or AMPK) may compete for the same sites or are situated at different serines and/or threonines on LXR. Furthermore, GlcNAcylation and phosphorylation of LXR might be affected by ligand binding, which has been shown for SUMOylation and acetylation of LXR (Venteclef et al., 2010; Lee et al., 2009). A study by Torra et al. (Torra et al., 2008) reported that Ser[198] phosphorylation of LXRα in RAW macrophages was induced by both synthetic and natural oxysterol LXR ligands and reduced by the RXR ligand 9-*cis*-retinoc acid. As such, we cannot exclude the possibility that LXR O-GlcNAcylation may be positively or negatively regulated by LXR and/or RXR ligands. From our *in vitro* GlcNAcylation results (Anthonisen et al., 2010) we believe that the major O-GlcNAc site(s) on LXRα and LXRβ resides in the N-terminal region containing the AF1 and DBD, indicating that O-GlcNAcylation occur independently of ligand. However, under hyperglycemic conditions, ligand binding may recruit OGT to LXR via CAs, possibly PGC1α as reported for FoxO1 (Housley et al., 2009). A more detailed mapping of the GlcNAc sites on LXR and site-directed mutagenesis as well as identification of coregulators of LXR under hyperglycemic conditions, are under way in our laboratory to elucidate the biological role of O-GlcNAc on LXR. A complete summary of putative mechanisms of glucose-signaling to LXR, ChREBP and lipogenesis is depicted in Figure 6.

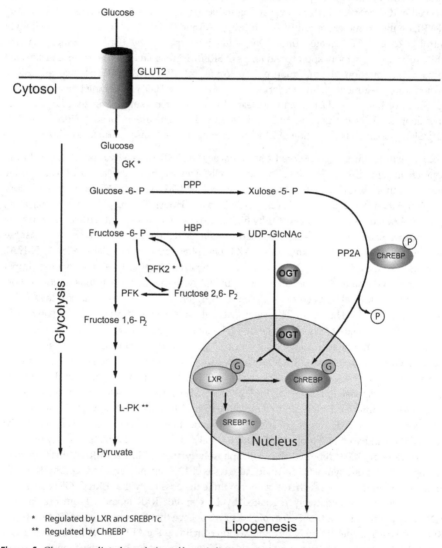

Figure 6. Glucose-mediated regulation of hepatic lipogenesis

5. Cross-talk between O-GlcNAc- and insulin signaling

Studies in *C.elegans* demonstrate that O-GlcNAc cycling phenotypes are very sensitive to insulin as well as nutrient composition and that levels of insulin and nutrients influence the role of O-GlcNAc cycling and vice versa (Mondoux et al., 2011; Hanover et al., 2010; Hanover et al., 2010; Whelan et al., 2008). Intriguingly, O-GlcNAc-marked promoters in *C.elegans* are biased toward genes associated with PIP3 signaling, hexosamine biosynthesis, and lipid/carbohydrate metabolism (Love et al., 2010a). Defects in O-GlcNAc cycling results in deregulation of genes necessary for carbohydrate and lipid metabolism in response to insulin (Forsythe et al., 2006; Hanover et al., 2010) suggesting that both O-GlcNAc cycling and insulin-signaling are required for a robust and adaptable response to hyperglycemia. Several studies have implicated O-GlcNAc cycling in the development of insulin resistance (reviewed in (Mondoux et al., 2011)). Mice overexpressing OGT in muscle or fat and mammalian cells overexpressing OGA develop insulin resistance (McClain, 2002; Arias et al., 2004; Vosseller et al., 2002). Later studies revealed that a subset of OGT was able to transiently translocate to the plasma membrane via association with PIP3 generated by insulin-activated PI3K (Yang et al., 2008). In response to increased glucose metabolism, PIP3-associated OGT can O-GlcNAcylate IR, IRS and Akt antagonizing insulin signaling (Yang et al., 2008; Whelan et al., 2010). Moreover, OGT may also interact with the mTOR pathway (Hanover et al., 2010). As mentioned in section 3.2.3, the downstream target for insulin signaling, FoxO1, is also modified by O-GlcNAc, apparently via OGT recruitment to PGC1α, providing another mechanism for OGT to contribute to insulin resistance, at least for sustained hepatic glucose production in response to hyperglycemia (Housley et al., 2009). Directing OGT to transcriptional targets implies that PGC1α can integrate multiple nutrient signals to regulate gene expression. Whether OGT via PGC1α or other CAs is also recruited to ChREBP- and LXR-regulated promoters is currently not known. OGT is recruited to and O-GlcNAcylate several coregulators and histone modifying enzymes (acetylases/deacetylases, metylases/demetylases) and even histones themselves (Fujiki et al., 2009; Hanover et al., 2012; Fujiki et al., 2011; Sakabe et al., 2010). Depending on the nutritional stimuli, all components of the transcriptional machinery from specific transcription factors to coregulators, histones and RNA polymerase II are subject to epigenetic regulation by acetylation, ubiquitinylation, SUMOylation, phosphorylation and/or O-GlcNAcylation (Rosenfeld et al., 2006; Venteclef et al., 2011; Love et al., 2010b; Kato et al., 2011). The fine-tuning of these modifications determines whether a gene is activated or repressed. Furthermore, as the substrate specificity of OGT is believed to be spatio-temporally regulated by transient interactions with large enzyme complexes, its binding to PIP3 may not occur solely at the plasma membrane, as PI3K is also active in the nucleus where it is involved in regulation of protein-chromatin interactions, transcription and mRNA export (Viiri et al., 2012; Kebede et al., 2012; Okada & Ye, 2009). As protein O-GlcNAcylation is rapidly increased at both the plasma membrane and the nucleus in response to serum-stimulation (Carrillo et al., 2011), OGT-binding to nuclear PIP3 may also be instrumental in transcriptional regulation in response to feeding. Interestingly, nonalcoholic fatty liver disease is often accompanied by hepatic insulin resistance, metabolic syndrome, and diabetes (reviewed in (Scorletti et al., 2011)) and the sensitivity of OGT to glucose increases with decreasing insulin signaling (Mondoux et al., 2011). These findings suggest that elevated O-GlcNAc cycling on key nuclear proteins contributes to the development of hepatic steatosis. This notion is also in line with the above mentioned observation by Guinez et al (Guinez et al., 2011), where overexpression of OGA reduced hepatic steatosis in db/db mice. A complete summary of a putative glucose-insulin cross-talk in regulation of hepatic *de novo* lipogenesis is depicted in Fig. 7.

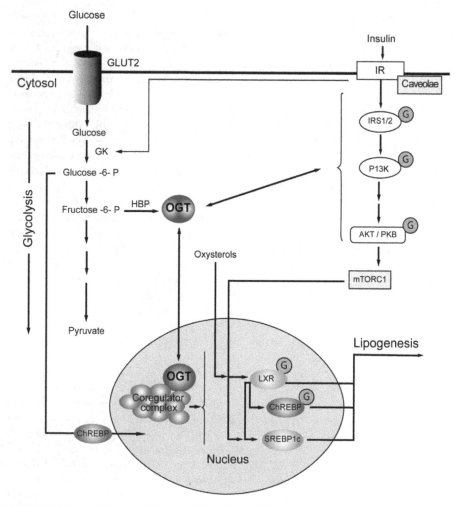

Figure 7. Glucose-insulin cross-talk in regulation of hepatic lipogenesis

6. Concluding remarks

In mice and humans, hepatic *de novo* lipogenesis is activated by a high intake of both glucose and fructose (Scorletti et al., 2011; Schwarz et al., 1995; Schwarz et al., 2003). Fructose increase hepatic hexosamine signaling (Hirahatake et al., 2011) and induce SREBP1c and ChREBP expression in hepatic cells (Matsuzaka et al., 2004; Haas et al., 2012; Koo et al., 2009), which may, in part be mediated by LXR. The response of LXR to glucose has been debated (Lazar & Willson, 2007), but a recent study support the notion of LXR as a glucose/fructose sensor as high sucrose fed mice exhibit elevated hepatic expression of SREBP1c and increased TG levels, which was not observed in LXRα/β double knock out mice (Korach-Andre et al., 2011). LXR increases lipogenesis, in part by activating SREBP1c and ChREBP proteins. Thus, in response to feeding, they can cooperatively activate most of the genes required for hepatic lipogenesis and TG secretion. Whether hepatic LXR drives the expression of SREBP1c and/or ChREBP to the same degree under different nutritional conditions is currently not known, as most studies have been performed using synthetic LXR agonists. We have preliminary results showing that hepatic expression of SREBP1c and ChREBP is upregulated in refed control mice and to a lesser extent in STZ-treated hyperglycemic mice, which is not observed in LXRα/β double knock out mice (Bindesbøll et al, unpublished). O-GlcNAc modification of LXR is increased in STZ-treated mice (Anthonisen et al., 2010) and we postulate that O-GlcNAc modification of LXR in response to glucose activates LXR and drives the expression of ChREBP and SREBP1c and in particular the lipogenic genes, ACC and SCD1. Furthermore, RNA Pol II ChIP-Seq data show reduced binding of RNA Pol II to the L-PK promoter and no binding of RNA Pol II to the SCD1 promoter in LXR α/β double knock out mice compared to control mice. Moreover, a novel LXRE immediately downstream of SCD1 was found, to which LXR bound more strongly than the previously published upstream LXR binding site (Boergesen et al., 2012). This suggests an important role for LXR as an upstream activator of ChREBP-mediated transcription and argues for LXR acting independently on the SCD1 promoter, at least under certain nutritional conditions. Previous studies have demonstrated that LXR directly activates key lipogenic genes (Joseph et al., 2002), most notably SCD1 in the liver of SREBP1c knockout mice (Liang et al., 2002; Chu et al., 2006). Why there would be a need for LXR to activate lipogenic genes directly, may be explained by the nutritional conditions and redundancy in the system. Oxysterols bind the endoplasmic reticulum resident Insig protein and could inhibit the proteolytic maturation of SREBP1c (Radhakrishnan et al., 2007). This would limit transcription by SREBP1c, and direct activation by LXR would be required to stimulate lipogenesis. In the absence of active SREBP1c, however, LXR may act in concert with ChREBP in regulating lipogenic expression. A recent study show that hepatic overexpression of ChREBP induces SCD1 expression and hepatic steatosis, but not insulin resistance (Benhamed et al., 2012). Whether overexpression of ChREBP affected LXR protein expression and transactivation of the SCD1 promoter was not investigated in this study. In later studies, it would be interesting to investigate the SCD1 expression and activity in livers or hepatocytes with targeted deletion of ChREBP. Benhamed et al (Benhamed et al., 2012) also showed that ChREBP expression was increased in liver biopsies from patients with steatosis and decreased in liver of patients with severe insulin resistance, suggesting that ChREBP, alone or in combination with LXR, drives SCD1 expression and steatosis independent of insulin resistance. This is in line with recent human studies showing no relationship between hepatic TG accumulation and insulin resistance (Cohen et al., 2011; Hooper et al., 2011). Thus, hepatic steatosis can either be the result or cause of hepatic insulin resistance. The mechanisms of hepatic insulin resistance is still not clear (Farese, Jr. et al., 2012), but may involve specific lipids, nutrition-induced metabolites and PTMs including O-GlcNAc. Hepatic TG synthesis may be a protective mechanism to limit accumulation of toxic free fatty acids, liver damage and fibrosis (Choi & Diehl, 2008) where particularly SCD1 seem to play a protective role (Li et al., 2009).

As LXR is shown also to act anti-inflammatory in liver (Wouters et al., 2008; Venteclef et al., 2010), LXR activation may be an important compensatory mechanism in response to excess nutrients to limit liver damage, inflammation and fibrosis. SUMOylation is an important ligand-activated transrepressional PTM of LXR on inflammatory genes (Venteclef et al., 2011) and future studies in our laboratory aim to elucidate a putative cross-talk between OGT and E3 ligases (SUMO conjugating enzymes) in liver in response to excess nutrients, especially high sugar levels (glucose and fructose). The relative roles of LXR, SREBP1c and ChREBP in driving lipogenesis is clearly dependent on both insulin and glucose signaling and cross-talk between these pathways. Both phosphorylation and GlcNAcylation appear instrumental in hepatic lipogenesis and future focus in our laboratory will be to elucidate a possible cross-talk between these PTMs, endogenous LXR ligands and interacting CAs in response to various feeding conditions (high glucose, fructose and/or fatty acids, cholesterol) and the impact on downstream ChREBP, SREBP1c and lipogenic enzyme expression and activity. ChIP and reChIP analysis in combination with loss of function studies have become powerful tools to analyze activation of specific genes by specific transcription factors in response to extracellular stimuli. By these methods, we anticipate that the signaling mechanisms and relative roles of LXR, ChREBP, SREBP1c and cooperating transcription factors in driving hepatic *de novo* lipogenesis will be revealed in the not too distant future.

Author details

Line M. Grønning-Wang, Christian Bindesbøll and Hilde I. Nebb
The Medical Faculty, Institute of Basic Medical Sciences,
Department of Nutrition, University of Oslo, Norway

7. References

Anthonisen, E.H., Berven, L., Holm, S., Nygard, M., Nebb, H.I., & Gronning-Wang, L.M. (2010). Nuclear receptor liver X receptor is O-GlcNAc-modified in response to glucose. *J.Biol.Chem.*, Vol. 285, No. 3, pp. 1607-1615, PM:19933273

Apfel, R., Benbrook, D., Lernhardt, E., Ortiz, M.A., Salbert, G., & Pfahl, M. (1994). A novel orphan receptor specific for a subset of thyroid hormone-responsive elements and its interaction with the retinoid/thyroid hormone receptor subfamily. *Mol.Cell Biol.*, Vol. 14, No. 10, pp. 7025-7035, PM:7935418

Arden, C., Tudhope, S.J., Petrie, J.L., Al-Oanzi, Z.H., Cullen, K.S., Lange, A.J., Towle, H.C., & Agius, L. (2012). Fructose 2,6-bisphosphate is essential for glucose-regulated gene transcription of glucose-6-phosphatase and other ChREBP target genes in hepatocytes. *Biochem.J.*, Vol. 443, No. 1, pp. 111-123, PM:22214556

Arias, E.B., Kim, J., & Cartee, G.D. (2004). Prolonged incubation in PUGNAc results in increased protein O-Linked glycosylation and insulin resistance in rat skeletal muscle. *Diabetes*, Vol. 53, No. 4, pp. 921-930, PM:15047606

Benhamed, F., Denechaud, P.D., Lemoine, M., Robichon, C., Moldes, M., Bertrand-Michel, J., Ratziu, V., Serfaty, L., Housset, C., Capeau, J., Girard, J., Guillou, H., & Postic, C. (2012). The lipogenic transcription factor ChREBP dissociates hepatic steatosis from insulin resistance in mice and humans. *J.Clin.Invest*, Vol. 122, No. 6, pp. 2176-2194, PM:22546860

Boergesen, M., Pedersen, T.A., Gross, B., van Heeringen, S.J., Hagenbeek, D., Bindesboll, C., Caron, S., Lalloyer, F., Steffensen, K.R., Nebb, H.I., Gustafsson, J.A., Stunnenberg, H.G., Staels, B., & Mandrup, S. (2012). Genome-wide profiling of liver X receptor, retinoid X receptor, and peroxisome proliferator-activated receptor alpha in mouse liver reveals extensive sharing of binding sites. *Mol.Cell Biol.*, Vol. 32, No. 4, pp. 852-867, PM:22158963

Bosco, D., Meda, P., & Iynedjian, P.B. (2000). Glucokinase and glucokinase regulatory protein: mutual dependence for nuclear localization. *Biochem.J.*, Vol. 348 Pt 1, No.pp. 215-222, PM:10794734

Brown, M.S., & Goldstein, J.L. (2008). Selective versus total insulin resistance: a pathogenic paradox. *Cell Metab*, Vol. 7, No. 2, pp. 95-96, PM:18249166

Calkin, A.C., & Tontonoz, P. (2010). Liver x receptor signaling pathways and atherosclerosis. *Arterioscler.Thromb.Vasc.Biol.*, Vol. 30, No. 8, pp. 1513-1518, PM:20631351

Carrillo, L.D., Froemming, J.A., & Mahal, L.K. (2011). Targeted in vivo O-GlcNAc sensors reveal discrete compartment-specific dynamics during signal transduction. *J.Biol.Chem.*, Vol. 286, No. 8, pp. 6650-6658, PM:21138847

Cha, J.Y., & Repa, J.J. (2007). The liver X receptor (LXR) and hepatic lipogenesis. The carbohydrate-response element-binding protein is a target gene of LXR. *J.Biol.Chem.*, Vol. 282, No. 1, pp. 743-751, PM:17107947

Chen, G., Liang, G., Ou, J., Goldstein, J.L., & Brown, M.S. (2004). Central role for liver X receptor in insulin-mediated activation of Srebp-1c transcription and stimulation of fatty acid synthesis in liver. *Proc.Natl.Acad.Sci.U.S.A*, Vol. 101, No. 31, pp. 11245-11250, PM:15266058

Chen, M., Bradley, M.N., Beaven, S.W., & Tontonoz, P. (2006). Phosphorylation of the liver X receptors. *FEBS Lett.*, Vol. 580, No. 20, pp. 4835-4841, PM:16904112

Chiang, J.Y., Kimmel, R., & Stroup, D. (2001). Regulation of cholesterol 7alpha-hydroxylase gene (CYP7A1) transcription by the liver orphan receptor (LXRalpha). *Gene,* Vol. 262, No. 1-2, pp. 257-265, PM:11179691

Choi, S.S., & Diehl, A.M. (2008). Hepatic triglyceride synthesis and nonalcoholic fatty liver disease. *Curr.Opin.Lipidol.,* Vol. 19, No. 3, pp. 295-300, PM:18460922

Chu, K., Miyazaki, M., Man, W.C., & Ntambi, J.M. (2006). Stearoyl-coenzyme A desaturase 1 deficiency protects against hypertriglyceridemia and increases plasma high-density lipoprotein cholesterol induced by liver X receptor activation. *Mol.Cell Biol.,* Vol. 26, No. 18, pp. 6786-6798, PM:16943421

Cohen, J.C., Horton, J.D., & Hobbs, H.H. (2011). Human fatty liver disease: old questions and new insights. *Science,* Vol. 332, No. 6037, pp. 1519-1523, PM:21700865

Costet, P., Luo, Y., Wang, N., & Tall, A.R. (2000). Sterol-dependent transactivation of the ABC1 promoter by the liver X receptor/retinoid X receptor. *J.Biol.Chem.,* Vol. 275, No. 36, pp. 28240-28245, PM:10858438

Denechaud, P.D., Bossard, P., Lobaccaro, J.M., Millatt, L., Staels, B., Girard, J., & Postic, C. (2008). ChREBP, but not LXRs, is required for the induction of glucose-regulated genes in mouse liver. *J.Clin.Invest,* Vol. 118, No. 3, pp. 956-964, PM:18292813

Deng, X., Zhang, W., Sullivan, I., Williams, J.B., Dong, Q., Park, E.A., Raghow, R., Unterman, T.G., & Elam, M.B. (2012). FoxO1 Inhibits Sterol Regulatory Element-binding Protein-1c (SREBP-1c) Gene Expression via Transcription Factors Sp1 and SREBP-1c. *J.Biol.Chem.,* Vol. 287, No. 24, pp. 20132-20143, PM:22511764

Dentin, R., Hedrick, S., Xie, J., Yates, J., III, & Montminy, M. (2008). Hepatic glucose sensing via the CREB coactivator CRTC2. *Science,* Vol. 319, No. 5868, pp. 1402-1405, PM:18323454

Dentin, R., Pegorier, J.P., Benhamed, F., Foufelle, F., Ferre, P., Fauveau, V., Magnuson, M.A., Girard, J., & Postic, C. (2004). Hepatic glucokinase is required for the synergistic action of ChREBP and SREBP-1c on glycolytic and lipogenic gene expression. *J.Biol.Chem.,* Vol. 279, No. 19, pp. 20314-20326, PM:14985368

Farese, R.V., Jr., Zechner, R., Newgard, C.B., & Walther, T.C. (2012). The problem of establishing relationships between hepatic steatosis and hepatic insulin resistance. *Cell Metab,* Vol. 15, No. 5, pp. 570-573, PM:22560209

Foretz, M., Pacot, C., Dugail, I., Lemarchand, P., Guichard, C., Le Liepvre, X., Berthelier-Lubrano, C., Spiegelman, B., Kim, J.B., Ferre, P., & Foufelle, F. (1999). ADD1/SREBP-1c is required in the activation of hepatic lipogenic gene expression by glucose. *Molecular and Cellular Biology,* Vol. 19, No. 5, pp. 3760-3768, ISI:000079821100052

Forsythe, M.E., Love, D.C., Lazarus, B.D., Kim, E.J., Prinz, W.A., Ashwell, G., Krause, M.W., & Hanover, J.A. (2006). Caenorhabditis elegans ortholog of a diabetes susceptibility locus: oga-1 (O-GlcNAcase) knockout impacts O-GlcNAc cycling, metabolism, and dauer. *Proc.Natl.Acad.Sci.U.S.A,* Vol. 103, No. 32, pp. 11952-11957, PM:16882729

Fujiki, R., Chikanishi, T., Hashiba, W., Ito, H., Takada, I., Roeder, R.G., Kitagawa, H., & Kato, S. (2009). GlcNAcylation of a histone methyltransferase in retinoic-acid-induced granulopoiesis. *Nature,* Vol. 459, No. 7245, pp. 455-459, PM:19377461

Fujiki, R., Hashiba, W., Sekine, H., Yokoyama, A., Chikanishi, T., Ito, S., Imai, Y., Kim, J., He, H.H., Igarashi, K., Kanno, J., Ohtake, F., Kitagawa, H., Roeder, R.G., Brown, M., & Kato, S. (2011). GlcNAcylation of histone H2B facilitates its monoubiquitination. *Nature,* Vol. 480, No. 7378, pp. 557-560, PM:22121020

Ghisletti, S., Huang, W., Ogawa, S., Pascual, G., Lin, M.E., Willson, T.M., Rosenfeld, M.G., & Glass, C.K. (2007). Parallel SUMOylation-dependent pathways mediate gene- and signal-specific transrepression by LXRs and PPARgamma. *Mol.Cell,* Vol. 25, No. 1, pp. 57-70, PM:17218271

Graf, G.A., Li, W.P., Gerard, R.D., Gelissen, I., White, A., Cohen, J.C., & Hobbs, H.H. (2002). Coexpression of ATP-binding cassette proteins ABCG5 and ABCG8 permits their transport to the apical surface. *J.Clin.Invest,* Vol. 110, No. 5, pp. 659-669, PM:12208867

Guertin, D.A., Stevens, D.M., Thoreen, C.C., Burds, A.A., Kalaany, N.Y., Moffat, J., Brown, M., Fitzgerald, K.J., & Sabatini, D.M. (2006). Ablation in mice of the mTORC components raptor, rictor, or mLST8 reveals that mTORC2 is required for signaling to Akt-FOXO and PKCalpha, but not S6K1. *Dev.Cell,* Vol. 11, No. 6, pp. 859-871, PM:17141160

Guinez, C., Filhoulaud, G., Rayah-Benhamed, F., Marmier, S., Dubuquoy, C., Dentin, R., Moldes, M., Burnol, A.F., Yang, X., Lefebvre, T., Girard, J., & Postic, C. (2011). O-GlcNAcylation increases ChREBP protein content and transcriptional activity in the liver. *Diabetes,* Vol. 60, No. 5, pp. 1399-1413, PM:21471514

Haas, J.T., Miao, J., Chanda, D., Wang, Y., Zhao, E., Haas, M.E., Hirschey, M., Vaitheesvaran, B., Farese, R.V., Jr., Kurland, I.J., Graham, M., Crooke, R., Foufelle, F., & Biddinger, S.B. (2012). Hepatic Insulin Signaling Is Required for Obesity-Dependent Expression of SREBP-1c mRNA but Not for Feeding-Dependent Expression. *Cell Metab,* Vol. 15, No. 6, pp. 873-884, PM:22682225

Hagiwara, A., Cornu, M., Cybulski, N., Polak, P., Betz, C., Trapani, F., Terracciano, L., Heim, M.H., Ruegg, M.A., & Hall, M.N. (2012). Hepatic mTORC2 activates glycolysis and lipogenesis through Akt, glucokinase, and SREBP1c. *Cell Metab,* Vol. 15, No. 5, pp. 725-738, PM:22521878

Han, C.Y., Umemoto, T., Omer, M., Den Hartigh, L.J., Chiba, T., LeBoeuf, R., Buller, C.L., Sweet, I.R., Pennathur, S., Abel, E.D., & Chait, A. (2012). NADPH oxidase-derived reactive oxygen species increases expression of monocyte chemotactic factor genes in cultured adipocytes. *J.Biol.Chem.,* Vol. 287, No. 13, pp. 10379-10393, PM:22287546

Hanover, J.A., Krause, M.W., & Love, D.C. (2010). The hexosamine signaling pathway: O-GlcNAc cycling in feast or famine. *Biochim.Biophys.Acta,* Vol. 1800, No. 2, pp. 80-95, PM:19647043

Hanover, J.A., Krause, M.W., & Love, D.C. (2012). Bittersweet memories: linking metabolism to epigenetics through O-GlcNAcylation. *Nat.Rev.Mol.Cell Biol.*, Vol. 13, No. 5, pp. 312-321, PM:22522719

Hart, G.W., Housley, M.P., & Slawson, C. (2007). Cycling of O-linked beta-N-acetylglucosamine on nucleocytoplasmic proteins. *Nature*, Vol. 446, No. 7139, pp. 1017-1022, PM:17460662

Hart, G.W., Slawson, C., Ramirez-Correa, G., & Lagerlof, O. (2011). Cross talk between O-GlcNAcylation and phosphorylation: roles in signaling, transcription, and chronic disease. *Annu.Rev.Biochem.*, Vol. 80, No.pp. 825-858, PM:21391816

Hasty, A.H., Shimano, H., Yahagi, N., Amemiya-Kudo, M., Perrey, S., Yoshikawa, T., Osuga, J., Okazaki, H., Tamura, Y., Iizuka, Y., Shionoiri, F., Ohashi, K., Harada, K., Gotoda, T., Nagai, R., Ishibashi, S., & Yamada, N. (2000). Sterol regulatory element-binding protein-1 is regulated by glucose at the transcriptional level. *J.Biol.Chem.*, Vol. 275, No. 40, pp. 31069-31077, PM:10913129

Havula, E., & Hietakangas, V. (2012). Glucose sensing by ChREBP/MondoA-Mlx transcription factors. *Semin.Cell Dev.Biol.*, PM:22406740

Heinz, S., Benner, C., Spann, N., Bertolino, E., Lin, Y.C., Laslo, P., Cheng, J.X., Murre, C., Singh, H., & Glass, C.K. (2010). Simple combinations of lineage-determining transcription factors prime cis-regulatory elements required for macrophage and B cell identities. *Mol.Cell*, Vol. 38, No. 4, pp. 576-589, PM:20513432

Herzog, B., Hallberg, M., Seth, A., Woods, A., White, R., & Parker, M.G. (2007). The nuclear receptor cofactor, receptor-interacting protein 140, is required for the regulation of hepatic lipid and glucose metabolism by liver X receptor. *Mol.Endocrinol.*, Vol. 21, No. 11, pp. 2687-2697, PM:17684114

Hirahatake, K.M., Meissen, J.K., Fiehn, O., & Adams, S.H. (2011). Comparative effects of fructose and glucose on lipogenic gene expression and intermediary metabolism in HepG2 liver cells. *PLoS.One.*, Vol. 6, No. 11, pp. e26583, PM:22096489

Hooper, A.J., Adams, L.A., & Burnett, J.R. (2011). Genetic determinants of hepatic steatosis in man. *J.Lipid Res.*, Vol. 52, No. 4, pp. 593-617, PM:21245030

Housley, M.P., Rodgers, J.T., Udeshi, N.D., Kelly, T.J., Shabanowitz, J., Hunt, D.F., Puigserver, P., & Hart, G.W. (2008). O-GlcNAc regulates FoxO activation in response to glucose. *J.Biol.Chem.*, Vol. 283, No. 24, pp. 16283-16292, PM:18420577

Housley, M.P., Udeshi, N.D., Rodgers, J.T., Shabanowitz, J., Puigserver, P., Hunt, D.F., & Hart, G.W. (2009). A PGC-1alpha-O-GlcNAc transferase complex regulates FoxO transcription factor activity in response to glucose. *J.Biol.Chem.*, Vol. 284, No. 8, pp. 5148-5157, PM:19103600

Huuskonen, J., Fielding, P.E., & Fielding, C.J. (2004). Role of p160 coactivator complex in the activation of liver X receptor. *Arterioscler.Thromb.Vasc.Biol.*, Vol. 24, No. 4, pp. 703-708, PM:14764426

Hwahng, S.H., Ki, S.H., Bae, E.J., Kim, H.E., & Kim, S.G. (2009). Role of adenosine monophosphate-activated protein kinase-p70 ribosomal S6 kinase-1 pathway in repression of liver X receptor-alpha-dependent lipogenic gene induction and hepatic steatosis by a novel class of dithiolethiones. *Hepatology*, Vol. 49, No. 6, pp. 1913-1925, PM:19378344

Ishii, S., Iizuka, K., Miller, B.C., & Uyeda, K. (2004). Carbohydrate response element binding protein directly promotes lipogenic enzyme gene transcription. *Proc.Natl.Acad.Sci.U.S.A*, Vol. 101, No. 44, pp. 15597-15602, PM:15496471

Jacinto, E., Facchinetti, V., Liu, D., Soto, N., Wei, S., Jung, S.Y., Huang, Q., Qin, J., & Su, B. (2006). SIN1/MIP1 maintains rictor-mTOR complex integrity and regulates Akt phosphorylation and substrate specificity. *Cell*, Vol. 127, No. 1, pp. 125-137, PM:16962653

Jakobsson, T., Venteclef, N., Toresson, G., Damdimopoulos, A.E., Ehrlund, A., Lou, X., Sanyal, S., Steffensen, K.R., Gustafsson, J.A., & Treuter, E. (2009). GPS2 is required for cholesterol efflux by triggering histone demethylation, LXR recruitment, and coregulator assembly at the ABCG1 locus. *Mol.Cell*, Vol. 34, No. 4, pp. 510-518, PM:19481530

Janowski, B.A., Willy, P.J., Devi, T.R., Falck, J.R., & Mangelsdorf, D.J. (1996). An oxysterol signalling pathway mediated by the nuclear receptor LXR alpha. *Nature*, Vol. 383, No. 6602, pp. 728-731, PM:8878485

Joseph, S.B., Laffitte, B.A., Patel, P.H., Watson, M.A., Matsukuma, K.E., Walczak, R., Collins, J.L., Osborne, T.F., & Tontonoz, P. (2002). Direct and indirect mechanisms for regulation of fatty acid synthase gene expression by liver X receptors. *J.Biol.Chem.*, Vol. 277, No. 13, pp. 11019-11025, PM:11790787

Kato, S., Yokoyama, A., & Fujiki, R. (2011). Nuclear receptor coregulators merge transcriptional coregulation with epigenetic regulation. *Trends Biochem.Sci.*, Vol. 36, No. 5, pp. 272-281, PM:21315607

Kebede, M., Ferdaoussi, M., Mancini, A., Alquier, T., Kulkarni, R.N., Walker, M.D., & Poitout, V. (2012). Glucose activates free fatty acid receptor 1 gene transcription via phosphatidylinositol-3-kinase-dependent O-GlcNAcylation of pancreas-duodenum homeobox-1. *Proc.Natl.Acad.Sci.U.S.A*, Vol. 109, No. 7, pp. 2376-2381, PM:22308370

Kim, K., Kim, K.H., Kim, H.H., & Cheong, J. (2008). Hepatitis B virus X protein induces lipogenic transcription factor SREBP1 and fatty acid synthase through the activation of nuclear receptor LXRalpha. *Biochem.J.*, Vol. 416, No. 2, pp. 219-230, PM:18782084

Kim, S.W., Park, K., Kwak, E., Choi, E., Lee, S., Ham, J., Kang, H., Kim, J.M., Hwang, S.Y., Kong, Y.Y., Lee, K., & Lee, J.W. (2003). Activating signal cointegrator 2 required for liver lipid metabolism mediated by liver X receptors in mice. *Mol.Cell Biol.*, Vol. 23, No. 10, pp. 3583-3592, PM:12724417

Kim, S.Y., Kim, H.I., Kim, T.H., Im, S.S., Park, S.K., Lee, I.K., Kim, K.S., & Ahn, Y.H. (2004). SREBP-1c mediates the insulin-dependent hepatic glucokinase expression. *J.Biol.Chem.*, Vol. 279, No. 29, pp. 30823-30829, PM:15123649

Kim, T.H., Kim, H., Park, J.M., Im, S.S., Bae, J.S., Kim, M.Y., Yoon, H.G., Cha, J.Y., Kim, K.S., & Ahn, Y.H. (2009). Interrelationship between liver X receptor alpha, sterol regulatory element-binding protein-1c, peroxisome proliferator-activated receptor gamma, and small heterodimer partner in the transcriptional regulation of glucokinase gene expression in liver. *J.Biol.Chem.*, Vol. 284, No. 22, pp. 15071-15083, PM:19366697

Koo, H.Y., Miyashita, M., Cho, B.H., & Nakamura, M.T. (2009). Replacing dietary glucose with fructose increases ChREBP activity and SREBP-1 protein in rat liver nucleus. *Biochem.Biophys.Res.Commun.*, Vol. 390, No. 2, pp. 285-289, PM:19799862

Korach-Andre, M., Archer, A., Gabbi, C., Barros, R.P., Pedrelli, M., Steffensen, K.R., Pettersson, A.T., Laurencikiene, J., Parini, P., & Gustafsson, J.A. (2011). Liver X receptors regulate de novo lipogenesis in a tissue-specific manner in C57BL/6 female mice. *Am.J.Physiol Endocrinol.Metab*, Vol. 301, No. 1, pp. E210-E222, PM:21521718

Kuo, M., Zilberfarb, V., Gangneux, N., Christeff, N., & Issad, T. (2008). O-glycosylation of FoxO1 increases its transcriptional activity towards the glucose 6-phosphatase gene. *FEBS Lett.*, Vol. 582, No. 5, pp. 829-834, PM:18280254

Lamming, D.W., Ye, L., Katajisto, P., Goncalves, M.D., Saitoh, M., Stevens, D.M., Davis, J.G., Salmon, A.B., Richardson, A., Ahima, R.S., Guertin, D.A., Sabatini, D.M., & Baur, J.A. (2012). Rapamycin-induced insulin resistance is mediated by mTORC2 loss and uncoupled from longevity. *Science*, Vol. 335, No. 6076, pp. 1638-1643, PM:22461615

Lazar, M.A., & Willson, T.M. (2007). Sweet dreams for LXR. *Cell Metab*, Vol. 5, No. 3, pp. 159-161, PM:17339022

Lee, J.H., Park, S.M., Kim, O.S., Lee, C.S., Woo, J.H., Park, S.J., Joe, E.H., & Jou, I. (2009). Differential SUMOylation of LXRalpha and LXRbeta mediates transrepression of STAT1 inflammatory signaling in IFN-gamma-stimulated brain astrocytes. *Mol.Cell*, Vol. 35, No. 6, pp. 806-817, PM:19782030

Lehmann, J.M., Kliewer, S.A., Moore, L.B., Smith-Oliver, T.A., Oliver, B.B., Su, J.L., Sundseth, S.S., Winegar, D.A., Blanchard, D.E., Spencer, T.A., & Willson, T.M. (1997). Activation of the nuclear receptor LXR by oxysterols defines a new hormone response pathway. *J.Biol.Chem.*, Vol. 272, No. 6, pp. 3137-3140, PM:9013544

Li, S., Brown, M.S., & Goldstein, J.L. (2010). Bifurcation of insulin signaling pathway in rat liver: mTORC1 required for stimulation of lipogenesis, but not inhibition of gluconeogenesis. *Proc.Natl.Acad.Sci.U.S.A*, Vol. 107, No. 8, pp. 3441-3446, PM:20133650

Li, X., Zhang, S., Blander, G., Tse, J.G., Krieger, M., & Guarente, L. (2007). SIRT1 deacetylates and positively regulates the nuclear receptor LXR. *Mol.Cell*, Vol. 28, No. 1, pp. 91-106, PM:17936707

Li, Z.Z., Berk, M., McIntyre, T.M., & Feldstein, A.E. (2009). Hepatic lipid partitioning and liver damage in nonalcoholic fatty liver disease: role of stearoyl-CoA desaturase. *J.Biol.Chem.*, Vol. 284, No. 9, pp. 5637-5644, PM:19119140

Liang, G.S., Yang, J., Horton, J.D., Hammer, R.E., Goldstein, J.L., & Brown, M.S. (2002). Diminished hepatic response to fasting/refeeding and liver X receptor agonists in mice with selective deficiency of sterol regulatory element-binding protein-1c. *Journal of Biological Chemistry*, Vol. 277, No. 11, pp. 9520-9528, ISI:000174400600105

Liu, X., Qiao, A., Ke, Y., Kong, X., Liang, J., Wang, R., Ouyang, X., Zuo, J., Chang, Y., & Fang, F. (2010). FoxO1 represses LXRalpha-mediated transcriptional activity of SREBP-1c promoter in HepG2 cells. *FEBS Lett.*, Vol. 584, No. 20, pp. 4330-4334, PM:20868688

Love, D.C., Ghosh, S., Mondoux, M.A., Fukushige, T., Wang, P., Wilson, M.A., Iser, W.B., Wolkow, C.A., Krause, M.W., & Hanover, J.A. (2010a). Dynamic O-GlcNAc cycling at promoters of Caenorhabditis elegans genes regulating longevity, stress, and immunity. *Proc.Natl.Acad.Sci.U.S.A,* Vol. 107, No. 16, pp. 7413-7418, PM:20368426

Love, D.C., Krause, M.W., & Hanover, J.A. (2010b). O-GlcNAc cycling: emerging roles in development and epigenetics. *Semin.Cell Dev.Biol.,* Vol. 21, No. 6, pp. 646-654, PM:20488252

Love, D.C.a.H.J.A. (2005). The Hexosamine Signaling Pathway: Deciphering the "O-GlcNAc Code". *Sci.STKE,* Vol. 312, No. re13, pp. 1-14,

Lu, M., Wan, M., Leavens, K.F., Chu, Q., Monks, B.R., Fernandez, S., Ahima, R.S., Ueki, K., Kahn, C.R., & Birnbaum, M.J. (2012). Insulin regulates liver metabolism in vivo in the absence of hepatic Akt and Foxo1. *Nat.Med.,* Vol. 18, No. 3, pp. 388-395, PM:22344295

Matsuda, T., Noguchi, T., Yamada, K., Takenaka, M., & Tanaka, T. (1990). Regulation of the gene expression of glucokinase and L-type pyruvate kinase in primary cultures of rat hepatocytes by hormones and carbohydrates. *J.Biochem.,* Vol. 108, No. 5, pp. 778-784, PM:1964454

Matsumoto, M., Han, S., Kitamura, T., & Accili, D. (2006). Dual role of transcription factor FoxO1 in controlling hepatic insulin sensitivity and lipid metabolism. *J.Clin.Invest,* Vol. 116, No. 9, pp. 2464-2472, PM:16906224

Matsuzaka, T., Shimano, H., Yahagi, N., Amemiya-Kudo, M., Okazaki, H., Tamura, Y., Iizuka, Y., Ohashi, K., Tomita, S., Sekiya, M., Hasty, A., Nakagawa, Y., Sone, H., Toyoshima, H., Ishibashi, S., Osuga, J., & Yamada, N. (2004). Insulin-independent induction of sterol regulatory element-binding protein-1c expression in the livers of streptozotocin-treated mice. *Diabetes,* Vol. 53, No. 3, pp. 560-569, PM:14988238

McClain, D.A. (2002). Hexosamines as mediators of nutrient sensing and regulation in diabetes. *Journal of Diabetes and Its Complications,* Vol. 16, No. 1, pp. 72-80, ISI:000174300400016

Mitro, N., Mak, P.A., Vargas, L., Godio, C., Hampton, E., Molteni, V., Kreusch, A., & Saez, E. (2007). The nuclear receptor LXR is a glucose sensor. *Nature,* Vol. 445, No. 7124, pp. 219-223, PM:17187055

Mondoux, M.A., Love, D.C., Ghosh, S.K., Fukushige, T., Bond, M., Weerasinghe, G.R., Hanover, J.A., & Krause, M.W. (2011). O-linked-N-acetylglucosamine cycling and insulin signaling are required for the glucose stress response in Caenorhabditis elegans. *Genetics,* Vol. 188, No. 2, pp. 369-382, PM:21441213

Oberkofler, H., Schraml, E., Krempler, F., & Patsch, W. (2003). Potentiation of liver X receptor transcriptional activity by peroxisome-proliferator-activated receptor gamma co-activator 1 alpha. *Biochem.J.,* Vol. 371, No. Pt 1, pp. 89-96, PM:12470296

Okada, M., & Ye, K. (2009). Nuclear phosphoinositide signaling regulates messenger RNA export. *RNA.Biol.,* Vol. 6, No. 1, pp. 12-16, PM:19106628

Ozcan, S., Andrali, S.S., & Cantrell, J.E. (2010). Modulation of transcription factor function by O-GlcNAc modification. *Biochim.Biophys.Acta,* Vol. 1799, No. 5-6, pp. 353-364, PM:20202486

Peet, D.J., Turley, S.D., Ma, W., Janowski, B.A., Lobaccaro, J.M., Hammer, R.E., & Mangelsdorf, D.J. (1998). Cholesterol and bile acid metabolism are impaired in mice lacking the nuclear oxysterol receptor LXR alpha. *Cell,* Vol. 93, No. 5, pp. 693-704, PM:9630215

Pehkonen, P., Welter-Stahl, L., Diwo, J., Ryynanen, J., Wienecke-Baldacchino, A., Heikkinen, S., Treuter, E., Steffensen, K.R., & Carlberg, C. (2012). Genome-wide landscape of liver X receptor chromatin binding and gene regulation in human macrophages. *BMC.Genomics,* Vol. 13, No.pp. 50, PM:22292898

Peterson, T.R., Sengupta, S.S., Harris, T.E., Carmack, A.E., Kang, S.A., Balderas, E., Guertin, D.A., Madden, K.L., Carpenter, A.E., Finck, B.N., & Sabatini, D.M. (2011). mTOR complex 1 regulates lipin 1 localization to control the SREBP pathway. *Cell,* Vol. 146, No. 3, pp. 408-420, PM:21816276

Radhakrishnan, A., Ikeda, Y., Kwon, H.J., Brown, M.S., & Goldstein, J.L. (2007). Sterol-regulated transport of SREBPs from endoplasmic reticulum to Golgi: oxysterols block transport by binding to Insig. *Proc.Natl.Acad.Sci.U.S.A,* Vol. 104, No. 16, pp. 6511-6518, PM:17428920

Repa, J.J., Berge, K.E., Pomajzl, C., Richardson, J.A., Hobbs, H., & Mangelsdorf, D.J. (2002). Regulation of ATP-binding cassette sterol transporters ABCG5 and ABCG8 by the liver X receptors alpha and beta. *J.Biol.Chem.,* Vol. 277, No. 21, pp. 18793-18800, PM:11901146

Repa, J.J., Liang, G., Ou, J., Bashmakov, Y., Lobaccaro, J.M., Shimomura, I., Shan, B., Brown, M.S., Goldstein, J.L., & Mangelsdorf, D.J. (2000). Regulation of mouse sterol regulatory element-binding protein-1c gene (SREBP-1c) by oxysterol receptors, LXRalpha and LXRbeta. *Genes Dev.,* Vol. 14, No. 22, pp. 2819-2830, PM:11090130

Rosenfeld, M.G., Lunyak, V.V., & Glass, C.K. (2006). Sensors and signals: a coactivator/corepressor/epigenetic code for integrating signal-dependent programs of transcriptional response. *Genes Dev.,* Vol. 20, No. 11, pp. 1405-1428, PM:16751179

Sabol, S.L., Brewer, H.B., Jr., & Santamarina-Fojo, S. (2005). The human ABCG1 gene: identification of LXR response elements that modulate expression in macrophages and liver. *J.Lipid Res.,* Vol. 46, No. 10, pp. 2151-2167, PM:16024918

Sakabe, K., Wang, Z., & Hart, G.W. (2010). Beta-N-acetylglucosamine (O-GlcNAc) is part of the histone code. *Proc.Natl.Acad.Sci.U.S.A,* Vol. 107, No. 46, pp. 19915-19920, PM:21045127

Saltiel, A.R., & Kahn, C.R. (2001). Insulin signalling and the regulation of glucose and lipid metabolism. *Nature,* Vol. 414, No. 6865, pp. 799-806, PM:11742412

Schwarz, J.M., Linfoot, P., Dare, D., & Aghajanian, K. (2003). Hepatic de novo lipogenesis in normoinsulinemic and hyperinsulinemic subjects consuming high-fat, low-carbohydrate and low-fat, high-carbohydrate isoenergetic diets. *Am.J.Clin.Nutr.*, Vol. 77, No. 1, pp. 43-50, PM:12499321

Schwarz, J.M., Neese, R.A., Turner, S., Dare, D., & Hellerstein, M.K. (1995). Short-term alterations in carbohydrate energy intake in humans. Striking effects on hepatic glucose production, de novo lipogenesis, lipolysis, and whole-body fuel selection. *J.Clin.Invest*, Vol. 96, No. 6, pp. 2735-2743, PM:8675642

Scorletti, E., Calder, P.C., & Byrne, C.D. (2011). Non-alcoholic fatty liver disease and cardiovascular risk: metabolic aspects and novel treatments. *Endocrine.*, Vol. 40, No. 3, pp. 332-343, PM:21894514

Shafi, R., Iyer, S.P., Ellies, L.G., O'Donnell, N., Marek, K.W., Chui, D., Hart, G.W., & Marth, J.D. (2000). The O-GlcNAc transferase gene resides on the X chromosome and is essential for embryonic stem cell viability and mouse ontogeny. *Proc.Natl.Acad.Sci.U.S.A*, Vol. 97, No. 11, pp. 5735-5739, PM:10801981

Shimano, H. (2001). Sterol regulatory element-binding proteins (SREBPs): transcriptional regulators of lipid synthetic genes. *Prog.Lipid Res.*, Vol. 40, No. 6, pp. 439-452, PM:11591434

Shimomura, I., Matsuda, M., Hammer, R.E., Bashmakov, Y., Brown, M.S., & Goldstein, J.L. (2000). Decreased IRS-2 and increased SREBP-1c lead to mixed insulin resistance and sensitivity in livers of lipodystrophic and ob/ob mice. *Mol.Cell*, Vol. 6, No. 1, pp. 77-86, PM:10949029

Soukas, A.A., Kane, E.A., Carr, C.E., Melo, J.A., & Ruvkun, G. (2009). Rictor/TORC2 regulates fat metabolism, feeding, growth, and life span in Caenorhabditis elegans. *Genes Dev.*, Vol. 23, No. 4, pp. 496-511, PM:19240135

Talukdar, S., & Hillgartner, F.B. (2006). The mechanism mediating the activation of acetyl-coenzyme A carboxylase-alpha gene transcription by the liver X receptor agonist T0-901317. *J.Lipid Res.*, Vol. 47, No. 11, pp. 2451-2461, PM:16931873

Teboul, M., Enmark, E., Li, Q., Wikstrom, A.C., Pelto-Huikko, M., & Gustafsson, J.A. (1995). OR-1, a member of the nuclear receptor superfamily that interacts with the 9-cis-retinoic acid receptor. *Proc.Natl.Acad.Sci.U.S.A*, Vol. 92, No. 6, pp. 2096-2100, PM:7892230

Tobin, K.A., Ulven, S.M., Schuster, G.U., Steineger, H.H., Andresen, S.M., Gustafsson, J.A., & Nebb, H.I. (2002). Liver X receptors as insulin-mediating factors in fatty acid and cholesterol biosynthesis. *J.Biol.Chem.*, Vol. 277, No. 12, pp. 10691-10697, PM:11781314

Torra, I.P., Ismaili, N., Feig, J.E., Xu, C.F., Cavasotto, C., Pancratov, R., Rogatsky, I., Neubert, T.A., Fisher, E.A., & Garabedian, M.J. (2008). Phosphorylation of liver X receptor alpha selectively regulates target gene expression in macrophages. *Mol.Cell Biol.*, Vol. 28, No. 8, pp. 2626-2636, PM:18250151

Tzivion, G., Dobson, M., & Ramakrishnan, G. (2011). FoxO transcription factors; Regulation by AKT and 14-3-3 proteins. *Biochim.Biophys.Acta*, Vol. 1813, No. 11, pp. 1938-1945, PM:21708191

Ulven, S.M., Dalen, K.T., Gustafsson, J.A., & Nebb, H.I. (2005). LXR is crucial in lipid metabolism. *Prostaglandins Leukot.Essent.Fatty Acids,* Vol. 73, No. 1, pp. 59-63, PM:15913974

Uyeda, K., & Repa, J.J. (2006). Carbohydrate response element binding protein, ChREBP, a transcription factor coupling hepatic glucose utilization and lipid synthesis. *Cell Metab,* Vol. 4, No. 2, pp. 107-110, PM:16890538

Venkateswaran, A., Laffitte, B.A., Joseph, S.B., Mak, P.A., Wilpitz, D.C., Edwards, P.A., & Tontonoz, P. (2000). Control of cellular cholesterol efflux by the nuclear oxysterol receptor LXR alpha. *Proc.Natl.Acad.Sci.U.S.A,* Vol. 97, No. 22, pp. 12097-12102, PM:11035776

Venteclef, N., Jakobsson, T., Ehrlund, A., Damdimopoulos, A., Mikkonen, L., Ellis, E., Nilsson, L.M., Parini, P., Janne, O.A., Gustafsson, J.A., Steffensen, K.R., & Treuter, E. (2010). GPS2-dependent corepressor/SUMO pathways govern anti-inflammatory actions of LRH-1 and LXRbeta in the hepatic acute phase response. *Genes Dev.,* Vol. 24, No. 4, pp. 381-395, PM:20159957

Venteclef, N., Jakobsson, T., Steffensen, K.R., & Treuter, E. (2011). Metabolic nuclear receptor signaling and the inflammatory acute phase response. *Trends Endocrinol.Metab,* Vol. 22, No. 8, pp. 333-343, PM:21646028

Viiri, K., Maki, M., & Lohi, O. (2012). Phosphoinositides as regulators of protein-chromatin interactions. *Sci.Signal.,* Vol. 5, No. 222, pp. e19, PM:22550339

Vosseller, K., Wells, L., Lane, M.D., & Hart, G.W. (2002). Elevated nucleocytoplasmic glycosylation by O-GlcNAc results in insulin resistance associated with defects in Akt activation in 3T3-L1 adipocytes. *Proc.Natl.Acad.Sci.U.S.A,* Vol. 99, No. 8, pp. 5313-5318, PM:11959983

Wagner, B.L., Valledor, A.F., Shao, G., Daige, C.L., Bischoff, E.D., Petrowski, M., Jepsen, K., Baek, S.H., Heyman, R.A., Rosenfeld, M.G., Schulman, I.G., & Glass, C.K. (2003). Promoter-specific roles for liver X receptor/corepressor complexes in the regulation of ABCA1 and SREBP1 gene expression. *Mol.Cell Biol.,* Vol. 23, No. 16, pp. 5780-5789, PM:12897148

Wan, M., Leavens, K.F., Saleh, D., Easton, R.M., Guertin, D.A., Peterson, T.R., Kaestner, K.H., Sabatini, D.M., & Birnbaum, M.J. (2011). Postprandial hepatic lipid metabolism requires signaling through Akt2 independent of the transcription factors FoxA2, FoxO1, and SREBP1c. *Cell Metab,* Vol. 14, No. 4, pp. 516-527, PM:21982711

Whelan, S.A., Dias, W.B., Thiruneelakantapillai, L., Lane, M.D., & Hart, G.W. (2010). Regulation of insulin receptor substrate 1 (IRS-1)/AKT kinase-mediated insulin signaling by O-Linked beta-N-acetylglucosamine in 3T3-L1 adipocytes. *J.Biol.Chem.,* Vol. 285, No. 8, pp. 5204-5211, PM:20018868

Whelan, S.A., Lane, M.D., & Hart, G.W. (2008). Regulation of the O-linked beta-N-acetylglucosamine transferase by insulin signaling. *J.Biol.Chem.,* Vol. 283, No. 31, pp. 21411-21417, PM:18519567

White, M.F. (2003). Insulin signaling in health and disease. *Science,* Vol. 302, No. 5651, pp. 1710-1711, PM:14657487

Willy, P.J., Umesono, K., Ong, E.S., Evans, R.M., Heyman, R.A., & Mangelsdorf, D.J. (1995). LXR, a nuclear receptor that defines a distinct retinoid response pathway. *Genes Dev.,* Vol. 9, No. 9, pp. 1033-1045, PM:7744246

Wong, R.H., & Sul, H.S. (2010). Insulin signaling in fatty acid and fat synthesis: a transcriptional perspective. *Curr.Opin.Pharmacol.,* Vol. 10, No. 6, pp. 684-691, PM:20817607

Wouters, K., van Gorp, P.J., Bieghs, V., Gijbels, M.J., Duimel, H., Lutjohann, D., Kerksiek, A., van, K.R., Maeda, N., Staels, B., van, B.M., Shiri-Sverdlov, R., & Hofker, M.H. (2008). Dietary cholesterol, rather than liver steatosis, leads to hepatic inflammation in hyperlipidemic mouse models of nonalcoholic steatohepatitis. *Hepatology,* Vol. 48, No. 2, pp. 474-486, PM:18666236

Wu, X., & Williams, K.J. (2012). NOX4 pathway as a source of selective insulin resistance and responsiveness. *Arterioscler.Thromb.Vasc.Biol.,* Vol. 32, No. 5, pp. 1236-1245, PM:22328777

Yamamoto, T., Shimano, H., Inoue, N., Nakagawa, Y., Matsuzaka, T., Takahashi, A., Yahagi, N., Sone, H., Suzuki, H., Toyoshima, H., & Yamada, N. (2007). Protein kinase A suppresses sterol regulatory element-binding protein-1C expression via phosphorylation of liver X receptor in the liver. *J.Biol.Chem.,* Vol. 282, No. 16, pp. 11687-11695, PM:17296605

Yamashita, H., Takenoshita, M., Sakurai, M., Bruick, R.K., Henzel, W.J., Shillinglaw, W., Arnot, D., & Uyeda, K. (2001). A glucose-responsive transcription factor that regulates carbohydrate metabolism in the liver. *Proc.Natl.Acad.Sci.U.S.A,* Vol. 98, No. 16, pp. 9116-9121, PM:11470916

Yang, X., Ongusaha, P.P., Miles, P.D., Havstad, J.C., Zhang, F., So, W.V., Kudlow, J.E., Michell, R.H., Olefsky, J.M., Field, S.J., & Evans, R.M. (2008). Phosphoinositide signalling links O-GlcNAc transferase to insulin resistance. *Nature,* Vol. 451, No. 7181, pp. 964-969, PM:18288188

Yecies, J.L., Zhang, H.H., Menon, S., Liu, S., Yecies, D., Lipovsky, A.I., Gorgun, C., Kwiatkowski, D.J., Hotamisligil, G.S., Lee, C.H., & Manning, B.D. (2011). Akt stimulates hepatic SREBP1c and lipogenesis through parallel mTORC1-dependent and independent pathways. *Cell Metab,* Vol. 14, No. 1, pp. 21-32, PM:21723501

Yoshikawa, T., Shimano, H., Yahagi, N., Ide, T., Amemiya-Kudo, M., Matsuzaka, T., Nakakuki, M., Tomita, S., Okazaki, H., Tamura, Y., Iizuka, Y., Ohashi, K., Takahashi, A., Sone, H., Osuga, J., Gotoda, T., Ishibashi, S., & Yamada, N. (2002). Polyunsaturated fatty acids suppress sterol regulatory element-binding protein 1c promoter activity by inhibition of liver X receptor (LXR) binding to LXR response elements. *Journal of Biological Chemistry,* Vol. 277, No. 3, pp. 1705-1711, ISI:000173421300013

Yu, L., York, J., von, B.K., Lutjohann, D., Cohen, J.C., & Hobbs, H.H. (2003). Stimulation of cholesterol excretion by the liver X receptor agonist requires ATP-binding cassette transporters G5 and G8. *J.Biol.Chem.*, Vol. 278, No. 18, pp. 15565-15570, PM:12601003

Zhao, L.F., Iwasaki, Y., Nishiyama, M., Taguchi, T., Tsugita, M., Okazaki, M., Nakayama, S., Kambayashi, M., Fujimoto, S., Hashimoto, K., Murao, K., & Terada, Y. (2012). Liver X receptor alpha is involved in the transcriptional regulation of the 6-phosphofructo-2-kinase/fructose-2,6-bisphosphatase gene. *Diabetes,* Vol. 61, No. 5, pp. 1062-1071, PM:22415873

Zoncu, R., Efeyan, A., & Sabatini, D.M. (2011). mTOR: from growth signal integration to cancer, diabetes and ageing. *Nat.Rev.Mol.Cell Biol.,* Vol. 12, No. 1, pp. 21-35, PM:21157483

Chapter 4

The Role of Copper as a Modifier of Lipid Metabolism

Jason L. Burkhead and Svetlana Lutsenko

Additional information is available at the end of the chapter

1. Introduction

1.1. Copper homeostasis in mammals

Dietary copper enters the body largely through the small intestine. Two membrane transporters are essential for this process. The high affinity copper uptake protein Ctr1 is responsible for making copper that enters via the apical membrane available in the cytosol for further utilization (1), whereas the copper-transporting ATPase ATP7A facilitates copper exit from the enterocytes into circulation (2) (Figure 1). Complete genetic inactivation of either transporter in experimental animals is embryonically lethal (3-5). However, partial inactivation or tissue specific inactivation of ATP7A or Ctr1, respectively, in either case is associated with copper accumulation in the intestine, impaired copper entry into the bloodstream, and severe copper deficiency in many organs and tissues (1). Copper deficiency, in turn, produces distinct metabolic changes that are discussed in detail in the following sections.

The majority of absorbed dietary copper is initially delivered to the liver. Hepatocytes utilize copper for their metabolic needs (such as respiration and radical defense); they also synthesize and secrete the major copper containing protein in serum, ceruloplasmin, and prevent copper overload in the body by exporting excess copper via the canalicular membrane into the bile (Figure 1). These two important functions of hepatocytes (the production of ceruloplasmin and the removal of excess copper) are performed by another transporter, the copper transporting ATPase ATP7B, which is homologous to ATP7A (6, 7). Inactivation of ATP7B in patients with Wilson's disease and in animal models is associated with marked copper overload in the liver and pathologic changes including marked lipid dysregulation in the liver and the serum (discussed in the later sections).

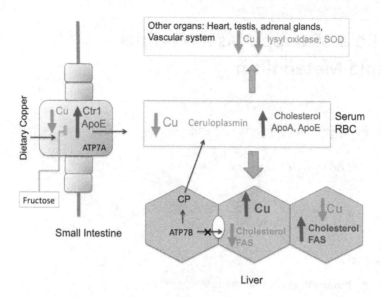

Figure 1. Copper homeostasis and lipid metabolism display functional interactions. Copper enters circulation via small intestine where copper transporters Ctr1 and ATP7A play the major role in the dietary copper absorption. Dietary copper deficiency is associated with a lower level of ceruloplasmin in the serum and increase in cholesterol and lipoproteins. Copper deficiency also upregulates the copper uptake protein Ctr1 in intestine; this compensatory effects is diminished by high fructose. In the liver, copper deficiency is associated with increased synthesis of cholesterol and higher expression of fatty acid synthase (FAS). Also, in the liver, the copper-transporting ATPase ATP7B mediates copper delivery to ceruloplasmin and copper export into the bile. Genetic inactivation of ATP7B in Wilson's disease is associated with copper overload and marked changes in lipid metabolism. Cholesterol biosynthesis is downregulated and both the serum and hepatic lipid profiles are altered.

The level of expression of copper transporters and their regulation varies between various organs (8). For example, ATP7B is the main transporter in hepatocytes, but it is absent from the adrenal gland, whereas opposite is true for ATP7A (9). Most of the tissues such as heart, brain, lung, placenta and kidneys express both copper-transporting ATPases along with the two copper uptake systems Ctr1 and Ctr2. As a result, the consequences of copper deficiency and copper overload are tissue specific, and certain organs such as heart or liver are more profoundly affected (see below). Recent studies also revealed an important homeostatic cross-talk between different organs in either copper deficiency or copper overload. For example, copper overload in the liver is accompanied by functional copper deficiency in the adrenal gland (9), whereas severe copper deficiency in the heart stimulates copper efflux from the liver (10), presumably to compensate for the deleterious effects of copper depletion.

The analysis of available literature also illustrates that variations in copper levels, either through the diet or as a result of genetic copper misbalance, have a profound effect on lipid metabolism. Significance of these observations is becoming more and more apparent given recent data that dietary influences (such as amount of fat in the diet) could be important modifiers of the course and severity of the disorders associated with copper misbalance (11). Reciprocally, copper deficiency has emerged as a factor in the development of Non-alcoholic Fatty-Liver Disease (NAFLD)(12, 13), although current reports paint complex picture (14) and further studies are needed. Despite ample phenotypic evidence in support of copper-lipid interactions, little mechanistic work has been done so far, and current understanding of this metabolic interaction at the molecular level is very limited. The goal of this review is to illustrate and emphasize the need for such detailed mechanistic investigations.

2. Copper deficiency

2.1. Copper in western diet

Copper deficiency has long been known to alter lipid metabolism; consequently, it has been proposed as a significant factor in human diseases associated with dyslipidemia (15). Copper deficiency is rarely diagnosed in humans, with a notable exception of a growing number of reports pointing to copper and other mineral insufficiencies as unintended consequences of bariatric surgeries (16-18). The under-detection of copper deficiency could be due to limitations of screening using serum or urine samples. Although liver is the main homeostatic organ for copper and has a high copper content, copper levels in serum and urine do not correlate well with a hepatic copper concentration (19), possibly masking deficiency in the liver.

Recent work using a categorical regression analysis of copper deficiency and excess shows a U-shaped dose-response curve. Compilation of data on toxicity due to copper excess and deficiency yielded a generalized linear model that was used to estimate adverse responses depending on copper dose or severity of copper limitation, as well as duration of copper misbalance (20). This model indicates that for humans the optimal intake level for Cu is 2.6 mg/day. The current United States Recommended Daily Intake is only 0.9 mg (US Food and Nutrition Board), whereas dietary study indicated that even 1.03 mg of Cu/day may be insufficient for adult men (21). The results of the third National Health and Nutrition Examination Survey (NHANES III, 2003) in the US showed that the mean daily intake of copper, depending on age, was 1.54–1.7 mg/day (±0.05 standard deviation (SD)) for men and 1.13-1.18 mg/d (±0.05 SD) for women. These results imply that a large portion of the population may have insufficient dietary copper intake and mild copper deficiency.

2.2. Low copper and human disease

Current data suggest that copper deficiency may be a common contributing factor in cardiovascular disease (CVD) and non-alcoholic fatty-liver disease (NAFLD) (22-24). As described above, surgical obesity treatment has also been strongly implicated in copper deficiency, likely by causing a diminished absorption of copper after a gastric bypass surgery (16-18). In addition, low copper levels were detected in organs, plasma and tissue of patients with several chronic diseases including cardiovascular disease, central nervous system, and musculoskeletal disorders (25). In fact, it was suggested that an ischemic heart disease could be largely attributed to copper deficiency (26), and that Cu deficiency and the high sugar consumption characteristic in the Western diet may interact in CVD (27). In a rat model, copper deficiency that reduces plasma levels of copper and ceruloplasmin also reduces copper content of the heart, liver, and testes. Coincidentally, heart zinc is also reduced, whereas hepatic iron levels rise (28), demonstrating the multifaceted effects of copper depletion on the overall body metal homeostasis.

Numerous studies in the rodent and other animal models (discussed below) provide strong indication of a significant link between copper deficiency and dyslipidemia. Specific studies linking human lipid response to copper deficiency are very limited. The available reports clearly illustrate the need for a better mechanistic insight. For example, studies using healthy male volunteers showed severe copper depletion with a diet containing 0.83 mg Cu/day, which is similar to levels in some contemporary diets. In these individuals, along with the diminished copper, the levels of serum copper-depended enzyme ceruloplasmin were reduced, as was copper-zinc superoxide dismutase activity in erythrocytes. In parallel, changes in lipid metabolism were evident. Cholesterol was elevated in the serum, and changes in the cholesterol levels were found to be more sensitive to copper levels than changes in hematology (29). These observations suggest that the dietary copper levels may be significant modifying factors in the disorders associated with lipid misbalance. Indeed, a role for copper deficiency is emerging in NAFLD (discussed below).

2.3. Low copper levels and hypercholesterolemia

Current data indicate that copper deficiency is associated with specific effects on systemic lipid metabolism. Although copper deficiency affects multiple organs (liver, heart, intestine, brain, adipose tissue), the liver and cardiovascular system appear more profoundly affected compared to other tissues. The effects of copper deficiency are partially reversible. In a middle-aged adult population with cardiovascular disease (CVD), copper supplementation was shown to raise the serum copper enzyme activities, but improvement of CVD measures was inconsistent (30). The same report also showed that copper supplementation reduced levels of oxidized serum LDL with statistical significance, but the results were inconsistent necessitating further research.

As described above, plasma cholesterol levels rise in human volunteers consuming a marginally low copper diet (29). In humans, the molecular mechanism behind this phenomenon has not yet been investigated in detail. However, much effort has been made characterizing the influence of copper deficiency on serum cholesterol as well as lipid profiles using animal models. These studies yielded wealth of useful information. Early work feeding rats a copper deficient diet revealed hypercholesterolemia, cardiac hypertrophy, hemorrhage, inflammation, and focal necrosis (31). This work also indicated that some of the cardiac pathology caused by copper deficiency could be linked to the lysyl oxidase deficiency. Lysyl oxidase (LOX) is a copper-dependent enzyme involved in the cross-linking of collagen and elastin. Aortas of copper-deficient rats showed deformation and loss of elasticity, though myocardial arteries were normal. Since lysyl oxidase is only one of the copper dependent enzymes required for proper heart functions (others include cytochrome C oxidase, cytosolic superoxide dismutase, dopamine beta-mono-oxygenase) the role of low copper in CVD may be realized through multiple mechanisms in addition to the altered lipid metabolism.

The development of cholesterolemia in response to copper deficiency has been explored in some detail in rats. Copper deficiency was found to be associated with the increased HDL and LDL levels (30% increase in HDL in one study). In addition, the plasma volume also increases, thus raising the available pool of cholesterol even more (22, 32). In effect, a 60% increase in total cholesterol may be observed (22). Overall, copper deficiency increases the absolute levels of cholesterol, but there is an argument whether the size of plasma cholesterol pool size may be a more powerful measure of cholesterol elevation. For example, the time course study of serum cholesterol in rats kept for 3-7 weeks on a copper deficient diet found that with time the plasma cholesterol concentration leveled off, but the pool size of cholesterol increased due to increases in the plasma volume (33). Curiously, the rates of HDL catabolism also increased, though the liver and the adrenal gland did not take up additional HDL in copper deficiency, suggesting one possible mechanism for dyslipidemia with respect to the HDL pool (34).

2.4. Serum lipoproteins: changes in structure, composition, and degradation

Rat models have proven to be especially consistent and valuable in studies of the dyslipidemia resulting from copper deficiency. Despite variations in the age, diet composition, and time on a copper deficient diet, an increase in the cholesterol concentration and/or pool was consistently reported in these animals. The lipids and lipoprotein components were analyzed in detail by al-Othman and co-workers and provided valuable data on changes in plasma pool size along with the composition and concentration of lipoproteins (33). These studies demonstrated no change in plasma phospholipid composition in copper deficient rats when compared to controls. In contrast, triglycerides, phospholipids, and cholesterol in LDL and HDL increased 2-fold or more. The VLDL composition of copper deficient animals changed most significantly with a 6-fold increase in triglycerides, 36% reduction in cholesterol and no change in phospholipid. An increase was observed in the VLDL particle size, but not the number of particles. In contrast, an increase in both size and number was seen for the LDL particles, whereas for the HDL particles an increase in the number of particles was observed, but no size change.

The composition of serum lipoproteins in copper deficient animals may also be influenced by shifts in the expression and distribution of apolipoproteins, linking physiological response in the liver to the observed changes in plasma cholesterol. The plasma levels of apolipoproteins A, B, and E increase in rats fed a copper deficient diet, consistent with the increased particle numbers and sizes (35-39). Remarkably, hepatic apoA1, ApoE and ApoB mRNA levels remain unchanged, suggesting that the alteration of the apolipoprotein levels is not due to an increase in transcription. Indeed, it was found that hepatic synthesis and secretion rates of apoA1 are upregulated (35), whereas the rates of synthesis of ApoB-48 and ApoB-100 are unchanged, despite an increase in secretion (37). Intestinal secretion of apoA-1 was also observed, providing yet another source for increased circulating apolipoproteins and cholesterol (40).

In copper deficiency, plasma HDL rich in apolipoprotein E (ApoE) accumulates and total ApoE binding to liver plasma membranes increases (also reported as a reduction in ApoE-free HDL binding). Interestingly, the cholesterol levels in the liver decrease with copper deficiency, despite an overall increase in hepatic cholesterol synthesis (41, 42). These changes in synthesis, binding properties, and redistribution of lipoproteins suggest some mechanisms through which copper deficiency affects serum cholesterol levels. Thus far, molecular investigations of copper deficiency have identified increased hepatic expression of SREBP1-responsive fatty-acid synthase, along with the increased nuclear localization of the mature SREBP1 transcriptional activator (24). Changes in the expression of other genes involved in the fatty acid synthesis have not been explored in any significant detail, although the mRNA levels for cholesterol 7-alpha hydroxylase were found decreased by 80%.

Changes in the plasma lipoprotein composition and structure are expected to influence the lipid content of red blood cells (RBC) through lipid exchange. Consequently, in copper deficiency changes in the RBC lipid composition are likely (43-45). Studies of the lipid profile of plasma and membranes of RBCs in copper deficiency support this view. Both cholesterol and phospholipid levels increase in the RBCs plasma membranes in Cu deficient rats, whereas the molar ratios of cholesterol:phospholipids and cholesterol:membrane protein are reduced (46). The phospholipid profiles change as well, with the increased stearic and docosadienoic acid content, and the lower levels of oleic and linolenic acid (47). A study assessing structural characteristics of the RBC membranes demonstrated a decrease in membrane fluidity and speculated that this could be the cause of hemolysis and anemia (48). Intriguingly, another study reported an increase in the RBC plasma membrane fluidity in copper deficient rats (49). This discrepancy may be due to experimental conditions. Motta and colleagues found that in copper deficiency, increased fluidity in RBCs plasma membrane can be seen alongside with higher rigidity due to enhanced susceptibility of triacylglycerol-rich lipoproteins to lipid peroxidation (50).

2.5. Copper-dependent lipid alteration in tissues

Given significant changes induced by copper deficiency in the serum and the liver, it is not surprising that the lipid and fatty-acid composition of other organs is also affected. Severe copper deficiency can be induced in C57BL mice by feeding dams a copper deficient diet and subsequently weaning pups to the same diet. These animals had lower levels of phospholipids in the liver and kidney, as well as lower triacylglycerols in kidneys (51). Lower proportion and total amount of di-homo-γ-linoleic acid was observed in all tissues of these mice, though levels of other lipids varied. Severe copper deficiency also induced hepatomegaly, reduced the brain weight, and reduced serum ceruloplasmin to 0.5% of control, indicating profound systemic effects.

The liver is an organ that experiences significant changes in lipid composition and membrane structure in response to copper deficiency. The loss of membrane fluidity in hepatic tissue has been reported and suggested to be caused by changes in the composition of unsaturated fatty acids and triacylglycerols of fatty acids (48). Other observed changes in membrane lipids associated with copper deficiency include a decreased ratio of monounsaturated:saturated C16 and C18 fatty acids in adipose tissue and a decreased fatty acid desaturase activity in liver microsomes (52). Phosphatidylcholine biosynthesis may also be affected, as choline phosphotransferase activity levels are lower both in the heart and liver tissue in copper deficient rats (53).

Concurrent with changes in lipid profiles and synthesis, copper deficiency decreases the total amounts of body fat and shifts metabolic fuel use from carbohydrate to fat. Respiratory quotient is reduced, but total energy intake is the same for animals kept on copper deficient and copper adequate diets (54). Young, weanling rats fed on a copper deficient diet for six weeks are leaner than controls, though they have increased serum cholesterol and triglycerides. Metabolically, whole body respiratory quotient decreases, reflected in a reduction of cardiac and adipose lipoprotein lipase, but not the skeletal muscle lipoprotein lipase (55). The change in fuel use may be related to upregulation of fatty acid synthesis. Copper deficiency does, however, decrease levels of hepatic cytochrome C oxidase (56).

The molecular mechanisms for these changes in lipid metabolism due to copper deficiency are understudied; as a result, current knowledge is limited. Increases in expression of specific apolipoproteins and increased transcription of gene for fatty acid synthase in the liver have been reported (see above), providing first insights into molecular players that are involved in response to low copper. Some key information has also been gained in one gene expression study. A transriptome analysis in the small intestine of copper deficient rats revealed upregulation of mRNA for proteins involved in cholesterol transport including apolipoprotein E and the lecithin:cholesterol acyltransferase providing mechanism for enhanced intestinal cholesterol secretion (57). The study also reported down-regulation of genes in the pathway for fatty acid beta-oxidation (both mitochondrial and peroxisomal). The results suggested a change in cell metabolism that reduced fatty acid oxidation, perhaps as a feedback to the decreased cytochrome C oxidase activity. The specific effects of copper deficiency may differ in tissues and serum. For example, in contrast to intestine, the activity of plasma lecithin:cholesterol acyltransferase (a risk factor for ischemic heart disease) is decreased in rats fed a copper deficient diet (58).

3. Functional interactions between copper, lipid, and other nutrients

3.1. Copper and dietary fat/lipid

Considering that copper deficiency influences systemic lipid metabolism, it would be interesting to know whether interactions between copper and lipid levels are reciprocal. In other words, it is important to determine whether dietary fat consumption, or changes in the type of fat consumed influences the activity or levels of copper-dependent enzymes. Changes in the ratios of saturated and unsaturated fatty acids have been noted in copper deficiency, and a cardioprotective effect of increasing proportion of polyunsaturated fatty acids was proposed (59). Curiously, feeding saturated fat in a copper deficient rat model increased hepatic copper as well as iron levels to a significant degree. Saturated fat consumption, however, did not change copper deficiency-induced lipid peroxidation, despite recovery of some hepatic copper. Copper-zinc superoxide dismutase (Cu/Zn-SOD) in the liver is less active in copper-deficient rats, whereas other hepatic antioxidant enzymes are unaffected by copper deficiency (56). This observation suggests that proper incorporation of copper in Cu/Zn-SOD may be key to preventing lipid peroxidation.

The effect of fatty acids on copper may also be mediated at the level of intestinal absorption. Experiments with the long chain fatty acids palmitate and stearate showed reduced levels of copper absorption from the jejunum (60) in response to treatment. In another study, direct cholesterol feeding of rabbits was used to model hypercholesterolemia and atherosclerosis. Adding cholesterol to 0.5% of diet triggered the redistribution of copper from the liver to plasma, with a 50% increase in plasma Cu and a 74% reduction in liver copper (61). Interestingly, copper supplementation in cholesterol-fed rabbits reduced atherosclerotic lesions (62). Further support for the importance of copper-lipid interactions in cardiac function is indicated by the observation that cardiomyopathy might be exacerbated by combination of high dietary fat and copper restriction. Specifically, when copper restriction and dietary fat supplementation were tested separately and together, the lowest level of cardiac cytochrome C oxidase activity was observed in copper-deficient rats on a high fat diet (63).

3.2. Fructose, lipids, and copper metabolism

The influence of dietary sugar consumption on lipid metabolism may be mediated, in part, by exacerbation of copper deficiency. Copper-deficient rats fed a sucrose-based or starch-based diets all had increased plasma cholesterol and lower plasma ceruloplasmin levels, as observed in copper deficiency alone (64). However, feeding sucrose rather than starch greatly enhanced deleterious effects of copper deficiency, such that those animals showed 60% mortality in the 9-week study. The copper deficient sucrose fed rats had a 3-fold lower hepatic copper level compared to starch-fed copper deficient rats. These results suggested a sucrose-dependent change in copper mobilization or retention within the liver (64). Cardiac abnormalities consistent with copper deficiency were also observed. Follow-up work indicated that when diets with fructose, glucose or starch were combined with copper deficiency, both glucose and fructose raised plasma cholesterol levels. However, severity of copper deficiency and mortality were much greater with fructose as opposed to glucose feeding (65), suggesting that the exacerbation of copper deficiency by sucrose was due to the fructose component. More recent work (66) indicates that the effect of fructose may be at the level of absorption, whereby copper deficiency induces upregulation of the copper transporter Ctr1, but this effect is eliminated by high fructose feeding (Figure 1).

Dietary fructose consumption and dietary copper deficiency independently alter fatty-acid metabolism; in combination, the effect is enhanced (66-69). Feeding weanling rats for three weeks on a diet adequate or deficient in copper along with either fructose or corn starch as a carbohydrate source (62% carbohydrate) revealed that fructose feeding enhanced indicators of copper deficiency, such as enhanced heart/body weight ratio and reduced hepatic copper (70). Analysis of hepatic enzymes involved in the lipid and carbohydrate metabolism also indicated that a diet deficient in copper had greatest metabolic effects in combination with fructose, less with glucose and least with starch (71). Glucose-6-phosphate dehydrogenase, malic enzyme, L-alpha-glycerophosphate dehydrogenase and fructose 1-6-diphosphatase were all unaffected by Cu deficiency, but their activities were enhanced most in combination with fructose, suggesting a complementary rather than direct role for fructose in exacerbating copper deficiency (71).

Copper deficiency in combination with fructose feeding alters the fatty acid composition of triacylglycerol in the heart and the liver. Cardiac phosphatidylinositol and phosphatidylserine were shown to increase nearly two-fold and arachidonic acid and docosapentaenoic acid to be elevated more than two-fold in copper deficient, fructose fed animals (70). A change in cardiac phospholipids may explain the increased mortality observed in copper deficient/fructose fed rats and suggest a potentially significant role for these dietary factors in ischemic heart disease. However, the observed changes in the lipid composition could not be correlated with the extent of copper deficiency, illustrating once again that a detailed mechanistic understanding of these dietary interactions could be highly beneficial.

Western diet is characterized by high fructose and high fat. In combination with a likely mild copper deficiency (see above), it stands to reason that these factors may all interact to induce changes in whole body metabolism, especially lipid metabolism, producing deleterious hepatic and cardiovascular effects. When this idea is tested in rats by combining high fat and fructose, or low fat and fructose with copper deficiency, liver metabolism is most significantly affected by the fat and fructose combination (28). Sugars mobilize copper from the liver to other tissues (64), possibly causing a change in liver physiology. Recent work also indicates that the mobilization of copper from the liver may be driven by heart-specific copper deficiency (10). Thus, in addition to interaction during absorption, an important inter-organ communication exists that jointly modulates levels of copper and lipids in various tissues

4. Copper misbalance and lipid metabolism in human disease

4.1. Lipid metabolism in Menkes disease

The studies discussed above were focused on the effects of dietary copper deficiency. In humans and in animals, genetic inactivation of the copper transporting ATPase ATP7A impairs copper export from the intestine, effectively limiting copper supply to many tissues and causing lethal pathology known as Menkes disease, MND (72). Depending on the type of mutation in ATP7A, severity of copper deficiency as well as disease manifestations vary between MND patients. Lipid profiles have been explored in some patients and found affected, but inconsistently, with the exception of a higher neutral lipid content of VLDL in all tested MND patients compared to controls. Interestingly, although ApoB in patients appeared normal, it degraded faster during storage suggesting lower stability (73). Variations in lipid profiles could also be age-related. Studies in animals (including an animal model for Menkes disease) demonstrate that unlike adults, copper deficient young animals do not show elevated cholesterol in the serum (74). The resistance of young mice and rats to hypercholesterolemia could be due to insufficient depletion of copper in the liver, which serves as a major store of copper in neonatal animals, or/and insufficiently progressed functional defects in the liver at young age (74).

ATP7A plays an important role in the development and maintenance of vasculature by supplying copper for functional maturation of lysyl oxidase. Recent studies also suggested the important role for ATP7A in the pathogenesis of atherosclerosis (75). ATP7A was detected in atherosclerotic lesions of mice with genetically inactivated LDL receptor where it colocalized with macrophages. Down-regulation of ATP7A in a macrophage-derived cell culture by siRNA resulted in decreased expression and enzymatic activity of cytosolic phospholipase A(2) alpha, an important enzyme involved in LDL oxidation (75). Furthermore, only cell-mediated LDL oxidation was reduced following down-regulation of ATP7A, whereas conditioned medium from either control or ATP7A down-regulated cells was without such effect (75). This result indicates that the reduced LDL oxidation is not simply due to a diminished copper export from cells with down-regulated ATP7A, but rather due to complex metabolic interactions between copper misbalance and lipid metabolism

4.2. Copper overload in Wilson's disease markedly alters lipid metabolism.

Direct evidence for the important role of copper in modulating hepatic lipid metabolism and serum lipid profiles has been produced by studies using *Atp7b-/-* mice (an animal model of Wilson's disease, WD) and, subsequently, samples from WD patients. Wilson's disease is a severe genetic disorder of copper overload, caused by inactivating mutations of copper transporting ATPase ATP7B and inability to excrete excess copper from hepatocytes (76). Despite massive accumulation of copper in a cytosol, copper incorporation into ceruloplasmin is impaired and serum levels and activity of ceruloplasmin are greatly diminished (phenotypically resembling consequences of copper deficiency, Figure 2). The genetically engineered knockouts of Atp7b (*Atp7b-/-* mice) accurately reproduce these key features of WD (77). Consequently, these animals have been used to investigate consequences of copper overload at the molecular level.

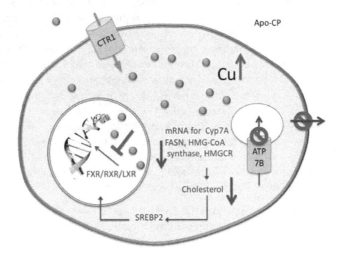

Figure 2. Elevated copper in *Atp7b-/-* hepatocytes inhibits activities of nuclear receptors and dysregulates lipid metabolism. In Wilson's disease, genetic inactivation of the copper transporting ATPase ATP7B results in an impaired copper export from hepatocytes, loss of copper incorporation into ceruloplasmin, CP, (and a secretion of Apo-CP into the serum) along with massive copper accumulation in hepatocytes. Copper concentrates preferentially in the cytosol and nuclei. Cytosolic copper does not prevent sensing of cholesterol levels, allowing SREBP-2 cleavage and entry into the nucleus in response to low cholesterol. However, high copper in the nucleus blocks activity of nuclear receptors, presumably through binding or oxidation. Impaired function of nuclear proteins is associated with a decrease in the mRNA levels for several enzymes, including HMG-CoA reductase (HMGCR) and HMG-CoA synthase, and low total cholesterol in the liver.

Studies at the early stage of pathology development in *Atp7b-/-* mice (prior to the onset of visible morphologic changes in the liver) were particularly informative. The analysis of liver transcriptome revealed that copper accumulation is associated with distinct metabolic changes: upregulation of genes involved in cell cycle and chromosome maintenance and down-regulation of lipid metabolism, especially cholesterol biosynthesis (78, 79). The mRNA studies were complemented by the analysis of metabolites, which confirmed significant (30-40%) decrease of cholesterol in the liver, and the lower levels of VLDL cholesterol in the serum (79). These early changes in lipid metabolism are maintained throughout the course of the disease as indicated by the analysis of the mRNA levels and serum lipid profiles at the later stages of the disease (80). Follow-up studies revealed that the sterol metabolism in a brain of young *Atp7b-/-* mice was also dysregulated, with cholesterol, 8-dehydrocholesterol, desmosterol, 7-dehydrocholesterol, and lathosterol all being highly increased (81, 82). It should be noted that the main cholesterol-sensing pathway that involves proteolysis and nuclear localization of SREBP2 transcription factor is not impaired in *Atp7b*[-/-] hepatocytes (79).

These important observations necessitated further mechanistic studies. Such work has recently been done using systems biology approach and biochemical/biophysical measurements. Direct *in situ* imaging of copper using X-fluorescence revealed the non-uniform distribution of accumulating copper in *Atp7b-/-* hepatocytes with a predominant increase in the cytosol and nuclei (80). Mass-spectrometry in combination with generation of protein networks provided strong evidence that accumulating copper had a significant and specific functional impact on *Atp7b-/-* nuclei, where it remodeled the RNA-processing machinery (83) and altered abundance and activity of nuclear receptors (84). Specifically, using quantitative Multidimensional Protein Identification Technology (MuDPIT), Wilmarth and colleagues found that the ligand-activated nuclear receptors FXR/NR1H4 and GR/NR3C1 were less abundant in nuclear preparations from *Atp7b-/-* liver, whereas the DNA repair machinery and the nucleus-localized glutathione peroxidase SelH were more abundant, consistent with the earlier transcriptome studies (79).

These findings provide important mechanistic insights into a copper-dependent dysregulation of lipid metabolism (summarized in Figure 2). It seems that hepatic nuclei are the primary sites of action for elevated copper. In WD, the local copper concentration in the nuclei can increase up to 50-100 fold (80). Copper is a redox active metal, and such marked elevation is likely to alter the nuclear redox environment, as also suggested by the upregulation of nuclear glutathione peroxidase SelH. Oxidation of sensitive cysteine residues and/or competition between copper and zinc in zinc fingers (which are common structural features in nuclear proteins) are likely mechanisms that impair activity of nuclear factors, such as transcription factors and/or components of the RNA splicing machinery (Figure 2). Current data suggest that neither potential cysteine oxidation nor copper-zinc competition are wide spread (83) and that nuclear proteins regulating lipid metabolism are preferentially affected by elevated copper. Thus, further studies are needed to understand the molecular basis of this increased sensitivity to copper levels.

The identification of FXR/LXR/RXR as important players in pathology development of WD (85) opens a new avenue for studies aimed on better understanding of the role of a diet (especially cholesterol and fat components) in the time-of-onset and severity of WD. It is important to emphasize that the effect of high copper on lipid metabolism is conserved between species, as evidenced by down-regulation of the same key enzymes in the lipid biosynthesis pathways in mice and human liver in response to copper overload (79). Recent studies of serum samples in the cohort of WD patient revealed differences in cholesterol metabolism that diminished with a copper-chelation therapy (82). These observations further supports high metabolic significance of copper:lipid interactions.

A marked effect of elevated copper on lipid metabolism was also reported in Long-Evans Cinnamon (LEC) rat, another murine model of WD (86). Analysis of the liver and serum lipid profiles in these animals showed that copper overload was associated with a higher content of triglycerides, free cholesterol and cholesteryl ester when compared to controls. The effect is actually opposite to that reported in *Atp7b-/-* mice. At present, the reason for this discrepancy is not clear. In LEC rat, liver disease is more severe compared to *Atp7b-/-* mice, invariably leading to animals death, which is not the case in mice. The mitochondrial function in these rats is more significantly affected (87), perhaps contributing to metabolic differences. Nevertheless, similarly to *Atp7b-/-* mice, the LEC serum is characterized by hypotriglyceridemia, hypocholesterolemia and abnormalities in the composition and size of circulating lipoproteins. Also similarly to *Atp7b-/-* and human liver, the activity of hepatic 3-hydroxy-3-methylglutaryl coenzyme A reductase is reduced in LEC rats, along with other important enzyme involved in lipid biosynthesis (Figure 2) (86).

4.3. Copper and NAFLD

The earlier studies of copper misbalance were focused mostly on the effects of copper deficiency and resulting dyslipidemia on cardiovascular disease. The liver, however, is the central organ of copper homeostasis and, as discussed above, it is greatly affected in WD. It may also be functionally affected in copper deficiency (88). Caloric excess associated with the modern Western diet is implicated in NAFLD, however caloric excess does not compensate for copper deficiency. Furthermore, copper deficiency can still be experienced by the liver, even when serum copper levels are maintained or increased due to factors such as dietary cholesterol.

Although numerous studies clearly link copper deficiency to altered lipid metabolism in animal models (12, 22, 24, 89-92) and human volunteers (29), only recently has low dietary copper been implicated in liver dyslipidemia pathology, including non-alcoholic fatty-liver disease (NAFLD) and non-alcoholic steatohepatitis (NASH). In a recent groundbreaking study, hepatic copper content in biopsy specimens was inversely correlated with the severity of fatty liver disease, and copper deficiency in a rodent model was found sufficient to induce NAFLD and metabolic syndrome (12). Hepatic iron accumulation, a known consequence of copper deficiency, is also observed in NAFLD (93). Iron accumulation likely results from the loss of holoceruloplasmin, a copper-dependent ferroxidase instrumental in iron distribution (94). This, in turn, results in lower levels of ferroportin in copper deficient rats; coincidentally, NAFLD patients show less ferroportin expression than controls (93). Copper supplementation has been suggested as a therapy for NAFLD based on study in which a diet-induced (high carbohydrate fat-free diet) NAFLD in rats was improved by treatment with a Cu(I)-nicotinate complex (95).

Studies of NAFLD also reflect the intersection of copper deficiency, hepatic lipid metabolism and consumption of fructose as causative agents in NAFLD. High dietary sugars, particularly fructose, have been implicated in development of NAFLD and NASH (96, 97). As discussed above, there is evidence that dietary fructose contributes to copper deficiency (66, 69), indicating cross talk between these dietary factors. Sucrose and fructose may have similar effects, as the enzyme sucrase acts in the digestive system to convert the disaccharide sucrose into fructose and glucose for transport in the bloodstream. High dietary fructose results in decreased CuZnSOD expression (98), lowering resistance of oxidative damage. A diet of 60% fructose affects lipid metabolism as well as antioxidant status, including CuZnSOD, in the liver of rats after 13 weeks of treatment (68). It is clear these high fructose diets induce NAFLD in rodent models, however these diets are typically 60-70% fructose and may not accurately reflect human fructose consumption (for review see (99)).

It is proposed that fructose metabolism induces oxidative stress, and this may trigger NAFLD. A copper-fructose feeding study indicated that lipid peroxidation due to copper deficiency and a 62% fructose diet could be reduced by supplementing vitamin E to 1 g/kg, however copper deficiency remained, indicating that copper deficiency with fructose feeding may not be entirely a result of oxidative stress (100). Nevertheless, hepatic lipid peroxidation is enhanced significantly in copper deficiency with fructose feeding, supporting the role for oxidative stress in liver disease through the impairment of hepatic antioxidant systems (66).

Tissue	Increase	Decrease	Other observation
Erythrocytes	cholesterol, phospholipid, stearic acid, docosadienoic acid	cholesterol:phospholipid ratio, cholesterol: mebmbrane protein ratio, linoleic acid, oleic acid	membrane fluidity?
Plasma and serum	triglycerides, phospholipids and cholesterol in lipoproteins; apoA, apoB, apoE; plasma volume	N/A	hematocrit
Liver	apolipoprotein secretion, cholesterol synthesis and secretion, fatty acid syntase expression	apoE-free HDL binding; monounsaturated:saturate d fatty acids ratio	no change in apolipoprotein transcript levels
Heart	stearic acid; docosahexaenoic acid; total phospholipid	elastic fibers; palmitic acid; oleic acid	hypertrophy; inflammation; distorted elastic fibers
Small intestine	transcripts in cholesterol transport	transcripts in fatty-acid beta-oxidation	NA
Kidney	NA	triacylglycrol and phospholipid	NA

Table 1. Tissue-specific effects of copper deficiency on lipid metabolism

5. Conclusions

Copper deficiency and copper overload have multiple and significant effects on systemic and cellular lipid metabolism. Recent studies indicate that copper misbalance is an emerging factor in dyslipidemia and/or fatty-liver disease. In turn, lipid metabolism could be an important modifier of the time-of-onset and severity in Wilson's disease. Work over several decades has yielded physiological and biochemical data on the consequences of copper deficiency, particularly with respect to lipid metabolism, in rodent models. Understanding of human disease would greatly benefit from further analysis of copper levels and lipid profiles in human clinical specimens, as well as assessment of the influence of other nutrients at the molecular level. Animal models will remain important. As much as has already been learned, further mechanistic studies are bound to yield molecular level understanding of the important copper:lipid relationship. It is still not clear how copper deficiency alters gene expression and protein expression/function to produce observed pathologies. Transcript profiling, proteomic analysis, and metabolite profiling, in both data-driven and targeted formats, promise to provide more mechanistic details in animal models that can be tested in human pathology.

Author details

Jason L. Burkhead
Department of Biological Sciences, University of Alaska Anchorage, Anchorage, AK, USA

Svetlana Lutsenko
Department of Physiology, Johns Hopkins University, Baltimore, MD, USA

6. References

[1] Nose Y, Kim BE, Thiele DJ. Ctr1 drives intestinal copper absorption and is essential for growth, iron metabolism, and neonatal cardiac function. Cell Metab. 2006 Sep;4(3):235-44.

[2] Vulpe CD, Packman S. Cellular copper transport. Annu Rev Nutr. 1995;15:293-322.

[3] Lee J, Prohaska JR, Thiele DJ. Essential role for mammalian copper transporter Ctr1 in copper homeostasis and embryonic development. Proc Natl Acad Sci U S A. 2001 Jun 5;98(12):6842-7.

[4] Kuo YM, Zhou B, Cosco D, Gitschier J. The copper transporter CTR1 provides an essential function in mammalian embryonic development. Proc Natl Acad Sci U S A. 2001 Jun 5;98(12):6836-41.

[5] Mercer JF, Ambrosini L, Horton S, Gazeas S, Grimes A. Animal models of Menkes disease. Adv Exp Med Biol. 1999;448:97-108.

[6] Lutsenko S, Gupta A, Burkhead JL, Zuzel V. Cellular multitasking: the dual role of human Cu-ATPases in cofactor delivery and intracellular copper balance. Arch Biochem Biophys. 2008 Aug 1;476(1):22-32.

[7] Gupta A, Lutsenko S. Human copper transporters: mechanism, role in human diseases and therapeutic potential. Future Med Chem. 2009 Sep;1(6):1125-42.

[8] Lutsenko S, Barnes NL, Bartee MY, Dmitriev OY. Function and regulation of human copper-transporting ATPases. Physiol Rev. 2007 Jul;87(3):1011-46.

[9] Gerbasi V, Lutsenko S, Lewis EJ. A mutation in the ATP7B copper transporter causes reduced dopamine beta-hydroxylase and norepinephrine in mouse adrenal. Neurochem Res. 2003 Jun;28(6):867-73.

[10] Kim B, Turski M, Nose Y, Casad M, Rockman H, Thiele D. Cardiac copper deficiency activates a systemic signaling mechanism that communicates with the copper acquisition and storage organs. Cell Metab. 2010 May 5;11(5):353-63.

[11] Kegley KM, Sellers MA, Ferber MJ, Johnson MW, Joelson DW, Shrestha R. Fulminant Wilson's disease requiring liver transplantation in one monozygotic twin despite identical genetic mutation. Am J Transplant. 2010 May;10(5):1325-9.

[12] Aigner E, Strasser M, Haufe H, Sonnweber T, Hohla F, Stadlmayr A, et al. A role for low hepatic copper concentrations in nonalcoholic Fatty liver disease. Am J Gastroenterol. 2010 Sep;105(9):1978-85.

[13] Lampon N, Tutor JC. Effect of valproic acid treatment on copper availability in adult epileptic patients. Clin Biochem. 2010 Sep;43(13-14):1074-8.

[14] Lampon N, Tutor JC. A preliminary investigation on the possible association between diminished copper availability and non-alcoholic fatty liver disease in epileptic patients treated with valproic acid. Ups J Med Sci. 2011 May;116(2):148-54.

[15] Stemmer KL, Petering HG, Murthy L, Finelli VN, Menden EE. Copper deficiency effects on cardiovascular system and lipid metabolism in the rat; the role of dietary proteins and excessive zinc. Ann Nutr Metab. 1985;29(6):332-47.

[16] Ernst B, Thurnheer M, Schultes B. Copper deficiency after gastric bypass surgery. Obesity (Silver Spring). 2009 Nov;17(11):1980-1.

[17] Gletsu-Miller N, Broderius M, Frediani J, Zhao V, Griffith D, Davis SJ, et al. Incidence and prevalence of copper deficiency following roux-en-y gastric bypass surgery. Int J Obes (Lond). 2012 Mar;36(3):328-35.

[18] Medeiros D. Gastric bypass and copper deficiency: a possible overlooked consequence. Obes Surg. 2011 Sep;21(9):1482-3; author reply 4-5.

[19] Evans J, Newman S, Sherlock S. Liver copper levels in intrahepatic cholestasis of childhood. Gastroenterology. 1978 Nov;75(5):875-8.

[20] Chambers A, Krewski D, Birkett N, Plunkett L, Hertzberg R, Danzeisen R, et al. An exposure-response curve for copper excess and deficiency. J Toxicol Environ Health B Crit Rev. 2010 Oct;13(7-8):546-78.

[21] Reiser S, Smith JJ, Mertz W, Holbrook J, Scholfield D, Powell A, et al. Indices of copper status in humans consuming a typical American diet containing either fructose or starch. Am J Clin Nutr. 1985 Aug;42(2):242-51.

[22] Lei K. Alterations in plasma lipid, lipoprotein and apolipoprotein concentrations in copper-deficient rats. J Nutr. 1983 Nov;113(11):2178-83.

[23] Saari J. Copper deficiency and cardiovascular disease: role of peroxidation, glycation, and nitration. Can J Physiol Pharmacol. 2000 Oct;78(10):848-55.

[24] Tang Z, Gasperkova D, Xu J, Baillie R, Lee J, Clarke S. Copper deficiency induces hepatic fatty acid synthase gene transcription in rats by increasing the nuclear content of mature sterol regulatory element binding protein 1. J Nutr. 2000 Dec;130(12):2915-21.

[25] Klevay L. Is the Western diet adequate in copper? J Trace Elem Med Biol. 2011 Oct 5.

[26] Klevay L. Ischemic heart disease as deficiency disease. Cell Mol Biol (Noisy-le-grand). 2004 Dec;50(8):877-84.

[27] Aliabadi H. A deleterious interaction between copper deficiency and sugar ingestion may be the missing link in heart disease. Med Hypotheses. 2008;70(6):1163-6.

[28] Wapnir R, Devas G. Copper deficiency: interaction with high-fructose and high-fat diets in rats. Am J Clin Nutr. 1995 Jan;61(1):105-10.

[29] Klevay L, Inman L, Johnson L, Lawler M, Mahalko J, Milne D, et al. Increased cholesterol in plasma in a young man during experimental copper depletion. Metabolism. 1984 Dec;33(12):1112-8.

[30] Disilvestro R, Joseph E, Zhang W, Raimo A, Kim Y. A randomized trial of copper supplementation effects on blood copper enzyme activities and parameters related to cardiovascular health. Metabolism. 2012 Mar 21.

[31] Allen K, Klevay L. Cholesterolemia and cardiovascular abnormalities in rats caused by copper deficiency. Atherosclerosis. 1978 Jan;29(1):81-93.

[32] Carr T, Lei K. High-density lipoprotein cholesteryl ester and protein catabolism in hypercholesterolemic rats induced by copper deficiency. Metabolism. 1990 May;39(5):518-24.

[33] al-Othman A, Rosenstein F, Lei K. Pool size and concentration of plasma cholesterol are increased and tissue copper levels are reduced during early stages of copper deficiency in rats. J Nutr. 1994 May;124(5):628-35.

[34] Carr T, Lei K. In vivo apoprotein catabolism of high density lipoproteins in copper-deficient, hypercholesterolemic rats. Proc Soc Exp Biol Med. 1989 Sep;191(4):370-6.

[35] Hoogeveen R, Reaves S, Lei K. Copper deficiency increases hepatic apolipoprotein A-I synthesis and secretion but does not alter hepatic total cellular apolipoprotein A-I mRNA abundance in rats. J Nutr. 1995 Dec;125(12):2935-44.

[36] Mazur A, Nassir F, Gueux E, Cardot P, Bellanger J, Lamand M, et al. The effect of dietary copper on rat plasma apolipoprotein B, E plasma levels, and apolipoprotein gene expression in liver and intestine. Biol Trace Elem Res. 1992 Aug;34(2):107-13.

[37] Nassir F, Mazur A, Serougne C, Gueux E, Rayssiguier Y. Hepatic apolipoprotein B synthesis in copper-deficient rats. FEBS Lett. 1993 May 3;322(1):33-6.

[38] Nassir F, Giannoni F, Mazur A, Rayssiguier Y, Davidson N. Increased hepatic synthesis and accumulation of plasma apolipoprotein B100 in copper-deficient rats does not result from modification in apolipoprotein B mRNA editing. Lipids. 1996 Apr;31(4):433-6.

[39] Norn M. Pemphigoid related to epinephrine treatment. Am J Ophthalmol: United States; 1977. p. 138.

[40] Wu J, Lei K. Copper deficiency increases total protein and apolipoprotein A-I synthesis in the rat small intestine. Proc Soc Exp Biol Med. 1996 Sep;212(4):369-77.

[41] Hassel C, Lei K, Carr T, Marchello J. Lipoprotein receptors in copper-deficient rats: apolipoprotein E-free high density lipoprotein binding to liver membranes. Metabolism. 1987 Nov;36(11):1054-62.

[42] Yount N, McNamara D, Al-Othman A, Lei K. The effect of copper deficiency on rat hepatic 3-hydroxy-3-methylglutaryl coenzyme A reductase activity. J Nutr Biochem. 1990 Jan;1(1):21-7.

[43] Gold J, Phillips M. Effects of membrane lipids and -proteins and cytoskeletal proteins on the kinetics of cholesterol exchange between high density lipoprotein and human red blood cells, ghosts and microvesicles. Biochim Biophys Acta. 1992 Oct 19;1111(1):103-10.

[44] Nikolic M, Stanic D, Baricevic I, Jones D, Nedic O, Niketic V. Efflux of cholesterol and phospholipids derived from the haemoglobin-lipid adduct in human red blood cells into plasma. Clin Biochem. 2007 Mar;40(5-6):305-9.

[45] Schick B, Schick P. Cholesterol exchange in platelets, erythrocytes and megakaryocytes. Biochim Biophys Acta. 1985 Feb 8;833(2):281-90.

[46] Jain S, Williams D. Copper deficiency anemia: altered red blood cell lipids and viscosity in rats. Am J Clin Nutr. 1988 Sep;48(3):637-40.

[47] Abu-Salah K, al-Othman A, Lei K. Lipid composition and fluidity of the erythrocyte membrane in copper-deficient rats. Br J Nutr. 1992 Sep;68(2):435-43.

[48] Lei K, Rosenstein F, Shi F, Hassel C, Carr T, Zhang J. Alterations in lipid composition and fluidity of liver plasma membranes in copper-deficient rats. Proc Soc Exp Biol Med. 1988 Jul;188(3):335-41.

[49] Rock E, Gueux E, Mazur A, Motta C, Rayssiguier Y. Anemia in copper-deficient rats: role of alterations in erythrocyte membrane fluidity and oxidative damage. Am J Physiol. 1995 Nov;269(5 Pt 1):C1245-9.

[50] Motta C, Gueux E, Mazur A, Rayssiguier Y. Lipid fluidity of triacylglycerol-rich lipoproteins isolated from copper-deficient rats. Br J Nutr. 1996 May;75(5):767-73.

[51] Cunnane S, McAdoo K, Prohaska J. Lipid and fatty acid composition of organs from copper-deficient mice. J Nutr. 1986 Jul;116(7):1248-56.

[52] Wahle E, Davies N. Effect of dietary copper deficiency. in the rat on fatty acid composition of adipose tissue and desaturase activity of liver microsomes. Br J Nutr. 1975 Jul;34(1):105-12.

[53] Cornatzer W, Haning J, Klevay L. The effect of copper deficiency on heart microsomal phosphatidylcholine biosynthesis and concentration. Int J Biochem. 1986;18(12):1083-7.

[54] Hoogeveen R, Reaves S, Reid P, Reid B, Lei K. Copper deficiency shifts energy substrate utilization from carbohydrate to fat and reduces fat mass in rats. J Nutr. 1994 Sep;124(9):1660-6.

[55] Wildman R, Mao S. Tissue-specific alterations in lipoprotein lipase activity in copper-deficient rats. Biol Trace Elem Res. 2001 Jun;80(3):221-9.

[56] Lynch S, Strain J. Dietary saturated or polyunsaturated fat and copper deficiency in the rat. Biol Trace Elem Res. 1989 Nov;22(2):131-9.

[57] Tosco A, Fontanella B, Danise R, Cicatiello L, Grober O, Ravo M, et al. Molecular bases of copper and iron deficiency-associated dyslipidemia: a microarray analysis of the rat intestinal transcriptome. Genes Nutr. 2010 Mar;5(1):1-8.

[58] Lau B, Klevay L. Plasma lecithin: cholesterol acyltransferase in copper-deficient rats. J Nutr. 1981 Oct;111(10):1698-703.

[59] Mozaffarian D, Micha R, Wallace S. Effects on coronary heart disease of increasing polyunsaturated fat in place of saturated fat: a systematic review and meta-analysis of randomized controlled trials. PLoS Med. 2010 Mar;7(3):e1000252.

[60] Wapnir R, Sia M. Copper intestinal absorption in the rat: effect of free fatty acids and triglycerides. Proc Soc Exp Biol Med. 1996 Apr;211(4):381-6.

[61] Klevay L. Dietary cholesterol lowers liver copper in rabbits. Biol Trace Elem Res. 1988 Jun;16(1):51-7.

[62] Lamb D, Reeves G, Taylor A, Ferns G. Dietary copper supplementation reduces atherosclerosis in the cholesterol-fed rabbit. Atherosclerosis. 1999 Sep;146(1):33-43.

[63] Jalili T, Medeiros D, Wildman R. Aspects of cardiomyopathy are exacerbated by elevated dietary fat in copper-restricted rats. J Nutr. 1996 Apr;126(4):807-16.

[64] Fields M, Ferretti R, Smith JJ, Reiser S. Effect of copper deficiency on metabolism and mortality in rats fed sucrose or starch diets. J Nutr. 1983 Jul;113(7):1335-45.

[65] Fields M, Ferretti R, Smith JJ, Reiser S. The interaction of type of dietary carbohydrates with copper deficiency. Am J Clin Nutr. 1984 Feb;39(2):289-95.

[66] Song M, Schuschke D, Zhou Z, Chen T, Pierce W, Wang R, et al. High Fructose Feeding Induces Copper Deficiency in Sprague-Dawley rats: A Novel Mechanism for Obesity Related Fatty Liver. J Hepatol. 2011 Jul 19.

[67] Fields M, Lewis C. Dietary fructose but not starch is responsible for hyperlipidemia associated with copper deficiency in rats: effect of high-fat diet. J Am Coll Nutr. 1999 Feb;18(1):83-7.

[68] Girard A, Madani S, Boukortt F, Cherkaoui-Malki M, Belleville J, Prost J. Fructose-enriched diet modifies antioxidant status and lipid metabolism in spontaneously hypertensive rats. Nutrition. 2006 Jul-Aug;22(7-8):758-66.

[69] Redman R, Fields M, Reiser S, Smith JJ. Dietary fructose exacerbates the cardiac abnormalities of copper deficiency in rats. Atherosclerosis. 1988 Dec;74(3):203-14.

[70] Cunnane S, Fields M, Lewis C. Dietary carbohydrate influences tissue fatty acid and lipid composition in the copper-deficient rat. Biol Trace Elem Res. 1989 -1990 Winter;23:77-87.

[71] Fields M, Ferretti R, Judge J, Smith J, Reiser S. Effects of different dietary carbohydrates on hepatic enzymes of copper-deficient rats. Proc Soc Exp Biol Med. 1985 Mar;178(3):362-6.

[72] Gasch AT, Kaler SG, Kaiser-Kupfer M. Menkes disease. Ophthalmology. 1999 Mar;106(3):442-3.

[73] Blackett PR, Lee DM, Donaldson DL, Fesmire JD, Chan WY, Holcombe JH, et al. Studies of lipids, lipoproteins, and apolipoproteins in Menkes' disease. Pediatr Res. 1984 Sep;18(9):864-70.

[74] Prohaska JR, Korte JJ, Bailey WR. Serum cholesterol levels are not elevated in young copper-deficient rats, mice or brindled mice. J Nutr. 1985 Dec;115(12):1702-7.

[75] Qin Z, Konaniah ES, Neltner B, Nemenoff RA, Hui DY, Weintraub NL. Participation of ATP7A in macrophage mediated oxidation of LDL. J Lipid Res. 2010 Jun;51(6):1471-7.

[76] Das SK, Ray K. Wilson's disease: an update. Nat Clin Pract Neurol. 2006 Sep;2(9):482-93.

[77] Lutsenko S. Atp7b-/- mice as a model for studies of Wilson's disease. Biochem Soc Trans. 2008 Dec;36(Pt 6):1233-8.

[78] Huster D, Lutsenko S. Wilson disease: not just a copper disorder. Analysis of a Wilson disease model demonstrates the link between copper and lipid metabolism. Mol Biosyst. 2007 Dec;3(12):816-24.

[79] Huster D, Purnat TD, Burkhead JL, Ralle M, Fiehn O, Stuckert F, et al. High copper selectively alters lipid metabolism and cell cycle machinery in the mouse model of Wilson disease. J Biol Chem. 2007 Mar 16;282(11):8343-55.

[80] Ralle M, Huster D, Vogt S, Schirrmeister W, Burkhead JL, Capps TR, et al. Wilson disease at a single cell level: intracellular copper trafficking activates compartment-specific responses in hepatocytes. J Biol Chem. 2010 Oct 1;285(40):30875-83.

[81] Sauer SW, Merle U, Opp S, Haas D, Hoffmann GF, Stremmel W, et al. Severe dysfunction of respiratory chain and cholesterol metabolism in Atp7b(-/-) mice as a model for Wilson disease. Biochim Biophys Acta. 2011 Dec;1812(12):1607-15.

[82] Seessle J, Gohdes A, Gotthardt DN, Pfeiffenberger J, Eckert N, Stremmel W, et al. Alterations of lipid metabolism in Wilson disease. Lipids Health Dis. 2011;10:83.

[83] Burkhead JL, Ralle M, Wilmarth P, David L, Lutsenko S. Elevated copper remodels hepatic RNA processing machinery in the mouse model of Wilson's disease. J Mol Biol. 2011 Feb 11;406(1):44-58.

[84] Wilmarth PA, Short KK, Fiehn O, Lutsenko S, David LL, Burkhead JL. A systems approach implicates nuclear receptor targeting in the Atp7b(-/-) mouse model of Wilson's disease. Metallomics. 2012 Jul 28;4(7):660-8.

[85] Burkhead JL, Gray LW, Lutsenko S. Systems biology approach to Wilson's disease. Biometals. 2011 Jun;24(3):455-66.

[86] Levy E, Brunet S, Alvarez F, Seidman E, Bouchard G, Escobar E, et al. Abnormal hepatobiliary and circulating lipid metabolism in the Long-Evans Cinnamon rat model of Wilson's disease. Life Sci. 2007 Mar 27;80(16):1472-83.

[87] Zischka H, Lichtmannegger J, Schmitt S, Jagemann N, Schulz S, Wartini D, et al. Liver mitochondrial membrane crosslinking and destruction in a rat model of Wilson disease. J Clin Invest. 2011 Apr;121(4):1508-18.

[88] Roberts E, Sarkar B. Liver as a key organ in the supply, storage, and excretion of copper. Am J Clin Nutr. 2008 Sep;88(3):851S-4S.

[89] al-Othman A, Rosenstein F, Lei K. Copper deficiency increases in vivo hepatic synthesis of fatty acids, triacylglycerols, and phospholipids in rats. Proc Soc Exp Biol Med. 1993 Oct;204(1):97-103.

[90] Cunnane S, Horrobin D, Manku M. Contrasting effects of low or high copper intake on rat tissue lipid essential fatty acid composition. Ann Nutr Metab. 1985;29(2):103-10.

[91] Petering H, Murthy L, O'Flaherty E. Influence of dietary copper and zinc on rat lipid metabolism. J Agric Food Chem. 1977 Sep-Oct;25(5):1105-9.

[92] Stemmer K, Petering H, Murthy L, Finelli V, Menden E. Copper deficiency effects on cardiovascular system and lipid metabolism in the rat; the role of dietary proteins and excessive zinc. Ann Nutr Metab. 1985;29(6):332-47.

[93] Aigner E, Theurl I, Haufe H, Seifert M, Hohla F, Scharinger L, et al. Copper availability contributes to iron perturbations in human nonalcoholic fatty liver disease. Gastroenterology. 2008 Aug;135(2):680-8.

[94] Miyajima H, Takahashi Y, Kono S. Aceruloplasminemia, an inherited disorder of iron metabolism. Biometals. 2003;16(1):205-13.

[95] Salama R, Nassar A, Nafady A, Mohamed H. A novel therapeutic drug (copper nicotinic acid complex) for non-alcoholic fatty liver. Liver Int. 2007 May;27(4):454-64.

[96] Alisi A, Manco M, Pezzullo M, Nobili V. Fructose at the center of necroinflammation and fibrosis in nonalcoholic steatohepatitis. Hepatology. 2011 Jan;53(1):372-3.

[97] Sanchez-Lozada L, Mu W, Roncal C, Sautin Y, Abdelmalek M, Reungjui S, et al. Comparison of free fructose and glucose to sucrose in the ability to cause fatty liver. Eur J Nutr. 2010 Feb;49(1):1-9.

[98] Francini F, Castro M, Schinella G, Garcia M, Maiztegui B, Raschia M, et al. Changes induced by a fructose-rich diet on hepatic metabolism and the antioxidant system. Life Sci. 2010 Jun 19;86(25-26):965-71.

[99] Yilmaz Y. Review article: fructose in non-alcoholic fatty liver disease. Aliment Pharmacol Ther. 2012 Apr 2.

[100] Silverman S, Fields M, Lewis C. The effect of vitamin E on lipid peroxidation in the copper-deficient rat. J Nutr Biochem. 1990 Feb;1(2):98-101.

Chapter 5

The 18 kDa Translocator Protein and Atherosclerosis in Mice Lacking Apolipoprotein E

Jasmina Dimitrova-Shumkovska, Leo Veenman, Inbar Roim and Moshe Gavish

Additional information is available at the end of the chapter

1. Introduction

1.1. Apolipoprotein E, inflammation and atherosclerosis

The inflammatory disease atherosclerosis is characterized by plaque formation in the cardiovascular system, which together with thrombosis can lead to obstruction of blood vessels, potentially leading to ischemia, stroke, and heart failure (Libby et al., 2009; Chen et al., 2010; Drake et al., 2011). Atherosclerosis is triggered and sustained by inflammation related cytokines, chemokines, adhesion molecules and by the cellular components of the immune system (Ross, 1999; Epstein et al., 2004). Cholesterol, most of it transported as a low density lipoprotein (LDL) particle in the bloodstream, supports foam cell formation in atherosclerotic plaques. In parallel, cholesterol plays an important role in steroidogenesis and bile production (Lacapere and Papadopoulos, 2003), which have been correlated with mitochondrial 18 kDa Translocator Protein (TSPO) and apolipoprotein E (apoE) expression (Fujimura et al., 2008; Gaemperli et al., 2011). Lipoproteins are lipid transport vehicles that ensure the solubility of lipids within aqueous biological environments. Apolipoproteins stabilize the surface of lipoproteins, serve as cofactors for enzymatic reactions, and present themselves as ligands for lipoprotein receptors. The soluble apolipoprotein gene family, which includes apoE, encodes proteins with amphipathic structures that allow them to exist at the water-lipid interface (Chan, 1989). ApoE is a polymorphic 229-aa, 34-kDa protein, which is present in the cell nucleus and cytosolic compartments (Mahley & Huang, 1999). The human gene, located on chromosome 19, encodes three alleles: apoE2 (frequency in the human population, 5–10%), apoE3 (60–70%), and apoE4 (15–20%). The isoforms differ only at residues 112 and 158 (Cedazo-Minguez & Cowburn, 2001). However, there is only one isoform of apoE in mouse and it behaves like human apoE3 (Strittmatter & Bova Hill, 2002). It is suggested that apoE deficiency in mice mimics the human apoE4 status, which implies reduced apoE3 levels relative to apoE4 levels (Buttini et al., 1999; Sheng et al., 1998).

ApoE is synthesized in several areas of the body, including the liver, where it is produced by hepatic parenchymal cells, and becomes a component in the surface of circulating triglyceride-rich lipoproteins [very low density lipoprotein (VLDL) and chylomicrons, or their remnants], and certain high density lipoprotein (HDL) particles (Mahley, 1988). ApoE plays a major role in the transport of lipids in the bloodstream, where it participates in the delivery and clearance of serum triglycerides, phospholipids, and cholesterol (Mahley, 1988). ApoE is also synthesized in the spleen, lungs, adrenals, ovaries, kidneys, muscle cells, and macrophages (Mahley, 1988). ApoE-containing lipoproteins are bound and internalized via receptor-mediated endocytosis by a number of proteins of the LDL receptor (LDLR) and LDLR-related protein (LRP) families (Davignon et al., 1998). ApoE is considered to be a ligand that binds to 27 clusters of negatively charged cysteine-rich repeats in the extracellular domains of all LDLR gene family members. It has been suggested that apoE made its entrance on the evolutionary stage long after the receptors to which it binds (Beffert et al., 2004). This also indicates that the original primordial functions of the LDLR family did not involve interactions with apoE. The original functions of the LDLR family may have been on the one hand transporting macromolecules between increasingly specialized cells and on the other hand serving as sensors for intercellular communication and environmental conditions (Beffert et al., 2003).

Cholesterol accumulation within atherosclerotic plaque occurs when cholesterol influx into the arterial wall (from apoB-containing lipoproteins) exceeds cholesterol efflux. Early in atherogenesis circulating monocytes are recruited to the arterial sub-endothelium where they differentiate into macrophages, ingest cholesterol, and develop into "foam cells" (Ross, 1973; 1999; Ross et al., 2001). Initially, monocytes adhere to activated endothelium on which up-regulated cell adhesion molecules (CAMs) are displayed, a dynamic process sensitive to inflammatory cytokines, shear stress, and oxidative insults (Chia, 1998). Induction of vascular cell adhesion molecule-1 (VCAM-1), a member of the immunoglobulin superfamily of CAMs, is increasingly described as the key factor in monocyte infiltration (Nakashima et al., 1998; Truskey et al., 1999). ApoE-knockout mice (apoE KO) have been extensively used to study the relation of hypercholesterolemia and lipoprotein oxidation to atherogenesis (Hoen et al., 2003; Yang et al., 2009; Kunitomo et al., 2009). ApoE-deficient mice have elevated VCAM-1 in aortic lesions (Nakashima et al., 1998), which enhances monocyte recruitment and adhesion (Ramos & Partridge, 2005), while apoE expression in the artery wall reduces early foam cell lesion formation (Hasty et al., 1999). These findings imply that apoE may influence early inflammatory responses by suppressing endothelial activation and CAM expression (Stannard et al., 2001). ApoE helps protect against atherosclerosis, in part by mediating hepatic clearance of remnant plasma lipoproteins (Weisgraber et al., 1994). When apoE is absent or dysfunctional, severe hyperlipidemia and atherosclerosis ensue (Kashyap et al., 1995; Linton & Fazio, 1999). ApoE is also abundant in atherosclerotic lesions, secreted by resident cholesterol-loaded macrophages (Linton & Fazio, 1999). This locally produced apoE is atheroprotective by contributing to reverse cholesterol transport and by inhibiting smooth muscle cell proliferation (Mahley et al., 1999; Mahley and Ji, 2006). ApoE exerts several functions regarding lipid and cholesterol transport and metabolism: 1) apoE functions as an important carrier protein in the redistribution of lipids among cells (by incorporation into HDL (as HDL-E); 2) it plays a prominent role in the transport of cholesterol (by incorporating into intestinally synthesized cholymicrons); and 3) it takes part metabolism of plasma cholesterol and triglyceride (by interaction with the LDLR and the receptor binding of apoE lipoproteins (Krul & Tikkanen, 1988; Quinn et al., 2004; Elliott et al., 2007).

ApoE has an established immune modulatory function in the peripheral immune response to bacteria and viruses (Mahley & Rall, 2000). It also modulates inflammatory responses in cell culture models *in vitro* and in *in vivo* models of brain injury, where an apoE mimetic therapeutic peptide has been shown to reduce CNS inflammation (Lynch et al., 2003; McAdoo et al., 2005; Aono et al., 2003). Involvement of apoE in injurious and inflammatory processes in the brain has attracted intensive attention (Drake et al., 2011; Potter and Wisniewski, 2012). In the brain, as well as in the cerebrospinal fluid, non-neuronal cell types, most notably astroglia and microglia, are the primary producers of apoE (Boyles et al., 1985; Quinn et al., 2004), while neurons preferentially express the receptors for apoE (Beffert et al., 2003). Regarding pathological conditions, it has been shown that human neuroblastoma cells, such as SK-N-SH, express apoE mRNA and apoE protein (Elliott et al., 2007). ApoE expression in neurons can be induced during nerve regeneration after injury and in growth and development of the CNS (Quinn et al., 2004). Moreover, Harris et al. (2004) showed that neuron-generated apoE tends to accumulate intracellularly, whereas astrocyte-generated apoE tends to be secreted. ApoE present in neurons is found in the cytoplasm (Han et al., 1994; Xu et al., 1996). The appearance of apoE in neurons may be due to neuronal synthesis under particular conditions, or by insertion into the cytoplasm of extracellular apoE (Dupont-Wallois et al., 1997). As neurons, human fibroblasts express low level of apoE under normal conditions, but under specific circumstances, such as apoptosis and nerve injury, they can produce increased levels of apoE (Do-Carmo et al., 2002; Quinn et al., 2004).

1.2. The 18kDa Translocator Protein (TSPO) and apolipoprotein E

Recent studies by us and others have indicated that the mitochondrial 18kDa Translocator Protein (TSPO), also known as peripheral-type benzodiazepine receptor (PBR) is present throughout the cardiovascular system and may be involved in cardiovascular disorders including atherosclerosis (Veenman and Gavish, 2006). The primary intracellular location of the TSPO is the outer mitochondrial membrane. Various studies over the course of the last 3 decennia have indicated that mitochondrial TSPO, potentially in relation to cardiovascular disease, is involved in the regulation of cholesterol transport into mitochondria in relation to bile production and steroidogenesis (Krueger and Papadopoulos, 1990; Papadopoulos et al., 2006). In particular, TSPO regulates cholesterol transport from the outer to the inner mitochondrial membrane which is the rate-limiting step in steroid and bile acid biosyntheses (Krueger and Papadopoulos, 1990; Lacapère and Papadopoulos, 2003; Veenman et al., 2007). Three-dimensional models of the channel formed by the five α-helices of the TSPO indicate that it can accommodate a cholesterol molecule in the space delineated by the five helices. According to these models, the inner surface of the channel formed by the TSPO molecule would present a hydrophilic but uncharged pathway, allowing amphiphilic cholesterol molecules to cross the outer mitochondrial membrane (Papadopoulos et al., 1997, 2006; Veenman et al., 2007). At cellular levels TSPO is present in virtually all of the cells of the cardiovascular system, where they appear to take part in the responses to various challenges that an organism and its cardiovascular system face (Veenman & Gavish, 2006), including atherosclerosis and accompanying symptoms (Onyimba et al., 2011; Bird et al., 2010; Dimitrova-Shumkovska et al., 2010a,b,c, 2012).

TSPO are located in various components of blood vessels, including endothelial cells where TSPO may take part in immunologic and inflammatory responses (Hollingsworth et al., 1985; Bono et al., 1999; Milner et al., 2004; Veenman & Gavish, 2006). To establish a factual correlation between atherogenic challenges and TSPO binding characteristics, we have previously assayed TSPO binding characteristics in different tissues of rats fed a high fat high cholesterol (HFHC) diet, in comparison to rats fed a normal diet (Dimitrova-Shumkovska et al., 2010a). It appeared that enhancement of oxidative stress in the aorta and liver due to the atherogenic HFHC diet was accompanied by significant reductions in TSPO binding density in these organs. Binding levels of the TSPO specific ligand [³H]PK 11195 in heart appeared not to be affected by the HFHC diet in this rat model.

Previous studies have shown that TSPO as well as apoE can be associated with processes such as: cholesterol metabolism, oxidative stress, apoptosis, glial activation, inflammation, and immune responses. As a ligand for cell-surface lipoprotein receptors, apoE can prevent atherosclerosis by clearing cholesterol-rich lipoproteins from plasma (Mahley and Huang, 1999). The TSPO protein has also been shown to be present in the plasma membrane of red blood cells, as well as in the plasma membrane of neutrofils, where it was shown to stimulate NADPH-oxidase activation of these cells. The plasma membrane forms of TSPO may be involved in heme metabolism, calcium channel modulation, cell growth, and immunomodulation. Furthermore, nucleus expulsion in mature erythrocytes is inhibited by excess cellular cholesterol (Fan et al., 2009). However, the involvement of the TSPO in this process has not been investigated. A recent study in cell culture showed that TSPO is important for the regulation of mitochondrial protoporphyrin IX and heme levels (Zeno et al., 2012). Thus, the TSPO appears to take part in various stages of red blood cell formation.

Furthermore, TSPO takes part in the regulation of gene expression for proteins involved in adhesion, which potentially may play a role in platelet aggregation (Bode et al., 2012; Veenman et al., 2012). ApoE has also been found to be involved in platelet aggregation, while TSPO platelet levels have been found to be increased with various neurological disorders, in particular stress related disorders (Veenman and Gavish, 2000, 2006, 2012). It has been suggested that platelet aggregation may be affected by nitric oxide (NO) generation via apoE, while other studies suggest that NO requires the TSPO to induce collapse of the mitochondrial membrane potential ($\Delta\Psi$m), mitochondrial reactive oxygen species (ROS) generation and cell death (Shargorodsky et al., 2012). Thus, the TSPO may present one pathway whereby NO does affect platelet aggregation. Furthermore, various alteration in TSPO density in the heart as a response to stress have been reported (Gavish et al., 1992; Veenman and Gavish, 2006), suggesting one aspect of involvement of TSPO in cardiovascular diseases, including cardiac ischemia. It has also been shown that apoE is involved in cardiac ischemia (Mahley, 1988).

Apparently as a consequence of its role in steroidogenesis, TSPO typically are very abundant in steroidogenic tissues (Benavides et al., 1983; De Souza et al., 1985). Steroid hormones can affect TSPO levels, while in turn TSPO provides a modulatory function for steroid hormone production by regulation of mitochondrial cholesterol transport (Veenman et al., 2007). It is known that cholesterol affects TSPO function (Falchi et al., 2007). Interestingly, apoE is also well expressed in steroidogenic organs such as adrenal gland, ovary, and testis (Blue et al., 1983; Elshourbagy et al., 1985; Law et al., 1997). Nonetheless, studies by us suggests that elevated cholesterol levels, such as found in apoE KO mice, do not appear to affect TSPO levels in steroidogenic organs (Inbar Roim, M.Sc. Thesis, Technion – Israel Institute of Technology, 2008), even though effects in the cardiovascular system can be observed (Dimitrova-Shumkovska et al., 2010a). As has been reported, TSPO levels can be regulated by steroid hormones, which may be part of an organism's response to stress and injury (Anholt et al., 1985; Weizman et al., 1992; Gavish & Weizman, 1997; Gavish et al., 1999; Veenman et al., 2007; Mazurika et al., 2009; Veenman and Gavish, 2012). This suggests that TSPO levels may be part of a feedback control system for steroid production (responding to alterations in steroid levels), rather than be regulated by a feed forward signal provided by cholesterol (i.e. TSPO levels in relation to steroidogenesis are not being regulated by cholesterol levels *in vivo*) (Veenman and Gavish, 2012).

1.2.1. Involvement of TSPO in inflammation

Various studies have shown the presence of TSPO in all cell types of the immune system, thus proposed functional roles of the TSPO included modulation of stress-induced immunosuppression and immune cell activity (Lenfant et al., 1985; Ruff et al., 1985; Bessier et al., 1992; Marchetti et al., 1996; Bono et al., 1999; Veenman & Gavish, 2006). TSPO are present in platelets, lymphocytes, and mononuclear cells, and are also found in the endothelium, the striated cardiac muscle, the vascular smooth muscles, and the mast cells of the cardiovascular system (Veenman & Gavish, 2006). TSPO in the cardiovascular system appears to play roles in several aspects of the immune response, such as phagocytosis and the secretion of interleukin-2, interleukin-3, and immunoglobulin A (Veenman & Gavish, 2006). Mast cells are considered to be important for immune response to pathogens (Marshall, 2004) and they have also been implicated in the regulation of thrombosis and inflammation and cardiovascular disease processes such as atherosclerosis as well as in neoplastic conditions (Wojta et al., 2003). Studies have shown that benzodiazepines' inhibition of serotonin release in mast cells could reduce blood brain barrier permeability, influence pain levels, and decrease vascular smooth muscle contractions (Veenman and Gavish, 2006). Benzodiazepines have been found to bind to specific receptors constituted by the TSPO on macrophages and to modulate *in vitro* their metabolic oxidative responsiveness (Lenfant et al., 1985). TSPO in the cardiovascular system also has been associated with the development of atherosclerosis (Camici et al., 2012). It for example has been suggested that reductions in TSPO levels may act as a protective mechanisms against the development of oxidative stress in aorta and liver (Dimitrova-Shumkovska et al., 2010a, b, c; 2012).

Anti-inflammatory properties of TSPO ligands have been demonstrated in various tissues. TSPO ligands have been shown to reduce inflammation in animal models of rheumatoid arthritis (Waterfield et al., 1999), carrageenan-induced pleurisy (Torres et al., 2000), and pulmonary inflammation (Bribes et al., 2003). Taupin et al. (1993) have also demonstrated *in vivo* that the synthetic TSPO ligand Ro5-4864 increases brain IL-1, IL-6 and TNF-α production after brain trauma. These cytokines are known to play a role in the inflammatory reaction to brain injury (Heumann et al., 1987). Interestingly, one study showed that PK 11195, but not Ro5-4864, could exert anti-inflammatory actions on mononuclear phagocytes, regulating the release of IL-1b (Klegeris et al., 2000). In addition, *in vivo* studies have shown that TSPO ligands can reduce the typical inflammatory response presented by reactive microglia and reactive astroglia resulting from brain trauma (Ryu et al., 2005; Veiga et al., 2005).

1.3. Animal models and strategies for atherosclerosis study

Atherosclerotic plaques may appear early in life and might progress into severe, symptomatic plaques many decades later, dependent on the coexistence of risk factors such as age, genetic background, gender, hypercholesterolemia, hypertension, smoking, diabetes, etc. (Ross, 1999; Whitman, 2004). Rupture of lipid-rich coronary plaques can trigger an atherothrombotic event and probably is the most important mechanism inducing acute coronary syndrome (ACS) (Vilahuer et al., 2011).

Plaque rupture presents a major factor in ischemic processes associated with atherosclerosis (Zhao et al., 2008; Cheng et al., 2009; Gaemerli et al., 2011). Plaque rupture in the human condition, including the cardiovascular processes and events leading up to it, presently is virtually inaccessible for research. Therefore, animal models have been developed to study atherosclerosis, including plaque rupture and thrombus formation, and also how to take measures to prevent these from happening. Nonetheless, more sophisticated models need to be developed and tested to be able to better mimic the human condition. This is so, as mice and rats, for example, do not develop atherosclerosis without genetic manipulation, because they have a lipid physiology that is radically different from that in humans, as most of the cholesterol is being transported in HDL-like particles (Whitman, 2004; Singh et al., 2009; Vilahur et al., 2011). Furthermore, all of the existing animal models, including biological and mechanical triggering of atherogenesis, e.g., the Watanabe heritable hyperlipidemic (WHHL) rabbit model, the apolipoprotein E (ApoE) mouse model, and the LDL-receptor mouse model) suffer the drawback of lacking an end-stage atherosclerosis that would show plaque rupture accompanied by platelet and fibrin-rich occlusive thrombus at the rupture site (Singh et al., 2009). Another restriction of current models for cardiovascular disorders is that most of the studies explore only male mice to avoid effects of estrogens to the extent of lesion development and diminishing LDL oxidation (Caligiuri et al., 1999; Yang et al., 2004). As cardiovascular disorders also occur in women, it would be valuable to also study female animal research subjects. Furthermore, it would give direction to research relating hormonal conditions to atherosclerosis.

Cholesterol lowering by diet is associated with a reduction in DNA damage, at least in animal models (Singh et al., 2009). In general, modification of atherosclerotic risk factors by lipid lowering therapies, cessation of smoking, weight loss, and improved glucose control reduces circulating markers of inflammation. These and other findings suggest that inflammation is a primary process for atherosclerosis (Ziccardi et al., 2002; Rodriguez-Moran et al., 2003). Although high dietary intake of the anti-oxidant vitamin E and C has been associated with reduced risk of cardiovascular disease (CVD), well powered clinical trials in atherosclerosis-related CVD have indicated that supplements with vitamin C or vitamin E alone do not provide sufficient benefit, in comparison to, for example, statins (Kunitomo et al., 2009). Furthermore, specific antioxidants scavenge or metabolize some, but not all of the relevant oxidized molecules (Stocker and Keaney, 2004). Stocker and Keaney (2005) conclude that whenever a physiological process goes unchecked in case of disease, treatment strategies cannot simply rely on scavenging ROS. Nonetheless, drugs that have been proven to alter plaque progression have also been shown to alter vascular oxidative stress. For example, 3-hydroxy-3-methylglutaryl coenzyme A reductase (HMGCoA) inhibitors (Statins) reduce NAD(P)H oxidase activation and superoxide production *in vitro*, in part because of their capability to inhibit the membrane translocation (and thus activity) of the small GTP-binding protein Rac-1, which is a regulatory component of vascular NAD(P)H oxidase activation (Costopoulos et al., 2008). In conclusion, it appears that beneficial therapeutic treatments to prevent atherosclerosis include lowering of lipid levels and also reduction of oxidative stress. However, restricting a treatment to only reduction of oxidative stress does not appear to generate sufficient beneficial effects to counteract atherosclerosis.

2. The effects of cholesterol challenges that result in atherogenesis on TSPO binding density in aorta and heart

As apoE deficiency may increase cholesterol levels and induce NO generation, which in turn may affect TSPO function, we were interested to study whether TSPO binding characteristics may be affected in heart and aorta of apoE-knockout (B6.129P2-apoE^{tm1} N11) mice, in comparison to their C57BL/6 background mice (i.e. wild type, WT). For the present study homogenates of whole heart organ and aorta segments (aortic arch and descendending aorta) were used. For this approach, it was taken into consideration that accumulation of proatherogenic lipid affects all cells types present into vascular wall, and the response of the entire tissue to the cholesterol exposure is relevant as an indication of vascular defense as a whole (Hoen et al., 2003). All procedures with the animals were in accordance with National Institutes of Health (USA) guidelines for the care and use of experimental animals (NIH publication No. 85-23, revised 1996), and the experimental protocol was reviewed and approved by the local ethics committee. The mice were housed in polycarbonate cages in a pathogen – free facility set on a 12h light-dark cycle and given *ad libitum* access to water and standard laboratory feed. Prior to the experimental procedures, the rats were fed a commercial standard pellet feed (Filpaso, 52.11, Skopje, Republic of Macedonia), named "standard feed" hereafter.

At 16 weeks of age, animals were randomized into experimental groups: i) Two control groups (WT mice, n = 10) and (apoE KO mice, n = 10), both these control groups received standard feed for a additional period of 10 weeks; ii) Two experimental groups receiving the same feed for the same 10 weeks but supplemented with 1% cholesterol (1% WT mice, n = 10) and (1% apoE KO mice, n= 10); and iii) Two experimental groups received the same feed for the same 10 weeks but supplemented with 3% cholesterol (3% WT mice, n = 10) and (3% apoE KO mice, n = 10). After these 10 weeks, animals were sacrificed by cardiac puncture, under ketamine/xylazine anaesthesia, followed by the appropriated storage until application or procedures required for assays of TSPO binding characteristics, ROS parameters, and histopathology, as described in detail previously (Dimitrova-Shumkovska et al., 2010 a, b, c, 2012). Tissue homogenates of aorta and heart were prepared for our various assays. For TSPO binding assays, tissue homogenates were prepared in 50 mM PBS on ice with a Kinematika Polytron (Luzerne, Switzerland), as described previously (Dimitrova-Shumkovska et al., 2010 a, b, c). To prepare homogenates for assays of oxidative stress parameters, we used an Ultrasonic Homogenizer (Cole-Parmer Instrument Co., Chicago, IL) as described previously (Dimitrova-Shumkovska et al., 2010 a, b, c). For advanced oxidation protein products (AOPPs, Witko-Sarsat et al., 1996), tissue homogenates were prepared in 50 mM PBS at + 4 ºC, as described previously (Dimitrova-Shumkovska et al., 2010 a, b, c). For the other assays of oxidative stress (see below), tissue homogenates were prepared in 1.12 % KCl at + 4 ºC, as described previously (Dimitrova-Shumkovska et al., 2010 a, b, c). These later parameters of oxidative injury included: lipid peroxidation products [TBARs] (Draper and Hadley, 1990); protein carbonylation, PC (Shacter, 2000); superoxide dismutase activity (SOD assay kit, RA20408, Fluka, Biochemika, Steinheim, Germany), glutathione (GSH assay kit CS0260, Sigma-Aldrich, Steinheim, Germany), glutathione reductase (GSSG-Red), GRSA 114K4000, Sigma-Aldrich, Steinheim, Germany], Finally, aortas were prepared for anatomical observation and histopathology as described previously (Dimitrova-Shumkovska et al., 2010 a, b, c).

Effects of cholesterol supplements to the apoE KO mice on plaque formation in the aorta are shown in Figure 1. No atherosclerotic formation was found in WT mice regardless of diet (Figure 1A). Control aortas of apoE KO mice having access to standard feed are characterized by the presence of thin fibrous tissue caps i.e. encapsulations of collagen rich fibrous tissue without a necrotic core that showed only superficial accumulation of foam cells (Figure 1B). Cholesterol diet accelerated atherosclerosis in apoE KO mice, increasing the total surface area of plaque formation significantly over the intimal area (Figure 1C) compared to apoE mice receiving standard feed. In 1% cholesterol fed apoE KO mice, expansion of the necrotic core presenting an important pathogenic process contributing to plaque vulnerability was observed in comparison to standard fed apoE mice (Figure 1C). After administration of 3% cholesterol diet to apoE KO mice even more advanced lesions have developed. Initial xanthoma formation, cartilage tissue, and calcified nodules with an underlying fibrocalcific plaque with minimal or absence of necrosis occurred (Figure 1D). Furthermore, plaques become more progressive and lesions show luminal stenosis with pathologic intimal thickening. These observations are in line with other research data, where plague rupture was seen in apoE KO mice especially when exposed to western type diet

(Davignon et al., 1999; Johnson et al., 2005). ApoE KO mice can also develop interplaque hemorrhage and features of plaque instability that are accelerated by feeding westernized diet (Rosenfeld et al., 2000). "Western type diets for mice" typically utilize just one ingredient (milk fat or lard) as the primary source of energy from fat.

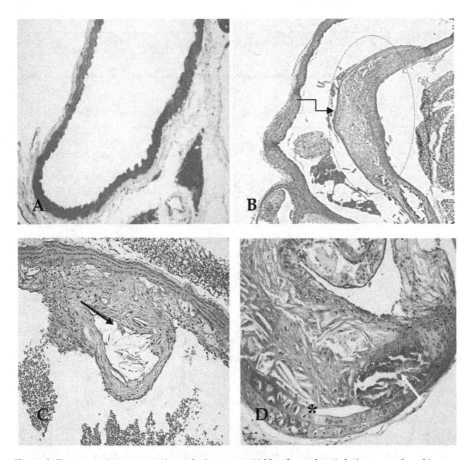

Figure 1. Representative cross-sections of mice aortas. A) No atherosclerotic lesions were found in wild-type mice regardless of the diet; B) atherosclerotic plaque (outlined) characterized by a thin fibrous tissue cap (elbow black arrow), particularly ssuperficial accumulation of foam cells (green arrow) without a necrotic core and encapsulated by collagen rich fibrous tissue in apoE KO mice given standard feed; C) accelerated atherosclerosis and deposition of cholesterol crystals (black arrow) in the endothelium of the aorta wall in 1% apoE KO; D) advanced lesions are developed in 3% apoE mice. Initial xanthoma formation, cartilage tissue (asterix) and calcified nodules (yellow arrow) with an underlying fibrocalcific plaque with minimal or absence of necrosis occur (H&E staining, microscopic magnification applied x 100).

Strain		Plasma lipoprotein levels mg/dL								
		Chol			TAG			HDL		
	wk	C	1%	3%	C	1%	3%	C	1%	3%
WT	16	67.7 ± 23.3	88.07 ± 28.0	110.0 ± 27.0**	71.0 ± 18.4	71.5 ± 5.1	97.5 ± 16.2**	33.2 ± 5.1	33.6 ± 10.8	74.7 ± 10.8***

Strain		Plasma lipoprotein levels mg/dL								
		Chol			TAG			HDL		
	wk	C	1%	3%	C	1%	3%	C	1%	3%
Apo E(-/-)	16	383.7 ± 47.3	457.23 ± 62*	555.4 ± 83.3***	117.7 ± 24.5	105.6 ± 11.4	159.4 ± 59.0**	67.0 ± 37.5	66.6 ± 16.3	32.9 ± 11.4**

Table 1. Effects of cholesterol (Chol) supplemented diet for 10 weeks, on lipoprotein levels in apoE KO mice and their WT counterparts. Unpaired Student t-*test* was performed. Data are expressed as mean ± SD; * = p < 0.05, ** = p < 0.01, *** = p < 0.001.

Changes in the serum levels of total cholesterol, triglycerides and HDL-cholesterol in each group are shown in **Table 1**. Corroborating previous studies (Davignon et al., 1999; Seo et al., 2005; Zhao et al., 2008) at 16 weeks of age, even before application of the cholesterol enriched diets, apoE KO mice, already displayed approximately 5 times higher levels of total cholesterol in comparison with WT mice. At this time point, no significant differences in triglycerides (TAG) levels were observed between WT mice and apoE KO mice. However, 3% diet regimes, caused significant increases in total cholesterol level in apoE KO mice (by 44%, p < 0.001), compared to standard feed. The enhanced total cholesterol levels, included an almost 90% representation of non HDL – cholesterol (calculated from Friedewald formula; Friedewald et al., 1972). In contrast, 3% WT mice, showed significantly higher cholesterol levels (by 62%, p < 0.01), including an almost 70% representation of HDL-lipoproteins. Supplement of 3% cholesterol also provoked significantly higher triglycerides levels: by 35 % (p < 0.01) in apoE mice and by 36% (p < 0.01) in WT mice. Supplement of 1% cholesterol, resulted in slight increases in total cholesterol in apoE mice (by 20%, p < 0.05), but did not significantly affect the triglycerides levels. The same type of diet did not affect lipoprotein levels in WT mice.

In the aorta, 3% cholesterol diet supplement, caused significant increases in "steady-state" levels of lipid peroxides (TBARs) and oxidized proteins in WT as well as apoE KO mice (**Table 2**). In detail, regarding lipid peroxidation, TBARs production was significantly increased by 2 fold in WT and apoE KO mice subjected to 3% cholesterol supplemented diet (+100%, p < 0.01 for WT mice, and +125%, p < 0.001 for ApoE KO mice). In parallel, protein oxidation products levels (AOPP) were also significantly higher (+135%, p < 0.01 in 3% WT mice and +177%, p < 0.001, in 3% apoE KO mice). Protein carbonyls (PC) showed a slight but non-significant increase in 3% cholesterol fed WT and apoE KO mice, compared to their controls. In contrast to the 3% diet regime, 1% cholesterol supplemented diet did not affect ROS parameters in aortic tissue in both WT and apoE KO mice.

Variables / Aorta	WT Control	1% Chol	3% Chol
TBARs nmol/mg	0.16 ± 0.04 (n=8)	0.17 ± 0.06 (n=7)	0.32 ± 0.08** (n=8)
AOPP nmol/mg	37.1 ± 11.3 (n=7)	44.1 ± 19.3 (n=8)	86.8 ± 21.4** (n=8)
PC pmol/mg	45.7 ± 11.0 (n=7)	55.2 ± 22.8 (n=8)	55.3 ± 11.8 (n=8)

Variables / Aorta	Apo E Control	1% Chol	3% Chol
TBARs nmol/mg	0.24 ± 0.07 (n=8)	0.29 ± 0.11 (n=8)	0.54 ± 0.25*** (n=10)
AOPP nmol/mg	22.0 ± 17.1 (n=8)	27.9 ± 7.1 (n=8)	60.5 ± 30.6*** (n=10)
PC pmol/mg	46.6 ± 20.3 (n=8)	45.3 ± 11.3 (n=8)	52.1 ± 10.6 (n=12)

Table 2. Effects of cholesterol (Chol) supplemented diet for 10 weeks on aorta oxidative stress parameters in apoE KO mice and their WT counterparts. 1-way ANOVA followed by application of the Tukey test to assess the significance of specific intergroup differences. Data are expressed as mean ± SD; * = $p < 0.05$, ** = $p < 0.01$, *** = $p < 0.001$.

The capacity of glutathione as an electron donor to regenerate the most important antioxidants (vitamin E, glutathione peroxidase (GPx), lipid hydroperoxides), is linked with the redox state of the glutathione disulfide – glutathione couple GSSG/2GSH (Schafer and Buettner, 2001). This in turn, has a high impact on the overall redox environment in the cell. Concerning antioxidant activities in aorta tissue due to 3% cholesterol supplemented feed, significantly reduced activity of superoxide dismutase (SOD) was measured in 3% apoE KO mice compared to standard feed mice (- 41%, **Table 3**). The results also suggest a significant reverse interaction between glutathione level (GSH) and glutathione peroxidase (GPx) activity in aorta tissue. In particular, the analyzed results indicated that the glutathione content in aorta of 3% apoE animals was significantly decreased (-32%), with simultaneous slight, but significant enhancement achieved in activity of glutathione peroxidase (+10%), as compared to standard feed control ($p < 0.05$). In parallel, glutathione content in aorta was also significantly reduced in 3% WT mice for 70% ($p < 0.01$), without affecting GPx levels. Feeding the mice diet supplemented with 1% cholesterol, resulted in significantly reduced activity in SOD in apoE KO mice (by 33% $p < 0.05$) and in WT mice (by 47% $p < 0.05$).

To determine TSPO binding characteristics in this paradigm we applied binding assays with the TSPO specific ligand [³H] PK 11195. The present study sought to determine whether cholesterol supplementation affects TSPO binding characteristics in aorta and heart of apoE KO mice in association with parameters for oxidative stress. Binding assays of the heart and

Variables / Aorta	WT Control	1% Chol	3% Chol
SOD U/mg	4.76 ± 1.5 (n=9)	2.5 ± 0.9* (n=7)	1.45 ± 0.6** (n=7)
GSH nmol/mg	7.8 ± 4.2 (n=9)	8.3 ± 3.7 (n=9)	3.7 ± 0.8** (n=9)
GPx mU/mg	0.261 ± 0.01 (n=8)	0.273 ± 0.01 (n=8)	0.257 ± 0.03 (n=8)

Variables / Aorta	Apo E Control	1% Chol	3% Chol
SOD U/mg	6.87 ± 1.6 (n=7)	4.6 ± 1.1* (n=6)	4.05 ± 0.88* (n=8)
GSH nmol/mg	6.7 ± 2.9 (n=9)	7.4 ± 2.7 (n=9)	4.5 ± 2.0* (n=9)
GPx mU/mg	0.242 ± 0.02 (n=9)	0.256 ± 0.01*(n=7)	0.266 ±0.03* (n=7)

Table 3. Effects of cholesterol (Chol) supplemented diet for 10 weeks on aorta antioxidant parameters in apoE KO mice and their WT counterparts. Unpaired Student t-*test* was performed. Data are expressed as mean ± SD; * = $p < 0.05$, ** = $p < 0.01$.

aorta with the TSPO specific ligand [³H]PK 11195 were done to determine potential effects of cholesterol supplementation on TSPO binding characteristics, according to methods described previously (Dimitrova-Shumkovska et al., 2010 a,b,c). For representative examples, see **Figure 2**. In heart , only in WT mice significant decreases in the B_{max} of TSPO (- 42%, p < 0.001) was determined with [³H]PK 11195 binding as a consequence of both cholesterol 1% and 3% supplemented diets, compared to control standard fed WT mice. Regarding the apoE KO mice, cholesterol supplemented diet did not induce differences in the TSPO binding characteristics in the heart **(Table 4)**. Regarding heart tissues, both in the apoE KO groups and WT groups, K_d values determined with [³H] PK 11195 binding were in the nM range (0.6 – 1.6 nM) showing no significant differences between experimental and control groups.

Regarding the aorta, feeding the mice with standard feed was not accompanied by significant differences in the TSPO binding characteristics of the aorta of apoE KO mice versus WT mice **(Table 4)**. Interestingly, these mouse aortas showed very TSPO binding levels, comparable to those observed in the adrenal of rats (Gavish et al., 1999). To date, the adrenal of rats is the tissue with one of highest demonstrated Bmax for TSPO ligand binding (Gavish et al., 1999). The 1% cholesterol supplemented diet significantly reduced TSPO binding capacity in aorta in both WT and apoE KO mice. In particular, reductions by 49% in WT mice and by 32% in apoE KO mice (p < 0.001 and p < 0.01, respectively) compared to their standard feed controls were observed **(Table 4)**. The 3% cholesterol diet also provoked a reduction in TSPO binding density by 58% in the aorta (p < 0.01), but only in WT mice. In the aortas of both groups, apoE KO mice and WT mice, K_d values determined with [³H] PK 11195 binding were in the nM range (1.5 – 2.6 nM), showing no significant differences between the groups.

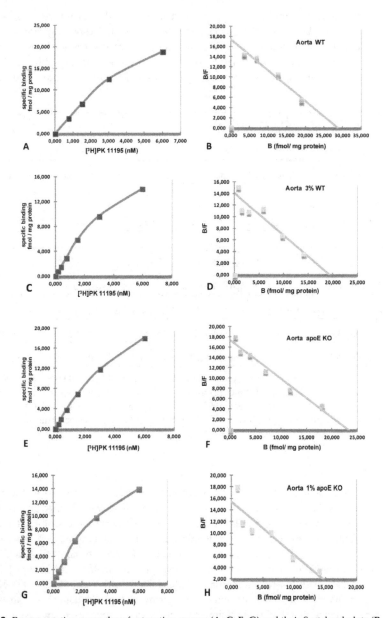

Figure 2. Representative examples of saturation curves (**A, C, E, G**) and their Scatchard plots (**B, D, F, H**) of [³H]PK 11195 binding to membrane homogenates of aorta, respectively of WT mice (**A , B, C, D**) and apoE KO mice (**E, F, G, H**). Abbreviations: apoE KO = apolipoprotein deficient mice; WT- wild type mice; B: bound; B/F: bound over free.

As the effects on TSPO binding density in heart and aorta due to intake of cholesterol supplemented diet take place primarily in the WT groups, and especially not in the 3% cholesterol diet fed apoE KO mice, these data suggest that decreases of TSPO binding density in heart and aorta may serve to counteract processes typically leading to cardiovascular damage, including atherosclerosis, as explained in more detail in the Discussion.

Bb - wild type mice								
	Bb - Control			Bb - 1% Chol.			Bb - 3 % Chol.	
Tissue	B max (fmol/mg)	Kd (nmol)	n	B max (fmol/mg)	Kd (nmol)	n	B max (fmol/mg)	Kd (nmol)	n
Heart	1740 ± 180	0.65 ± 0.1	5	1005 ± 240**	1.12 ± 0.3	6	1006 ± 140 **	1.32 ± 0.5	5
Aorta	29 000 ± 9700	2.50 ± 1.2	9	14 900 ± 3370***	2.31 ± 0.8	6	12 200 ± 2920**	2.62 ± 0.5	5

Apo E KO mice								
	Apo E - Control			Apo E - 1% Chol.			Apo E - 3 % Chol.	
Tissue	B max (fmol/mg)	Kd (nmol)	n	B max (fmol/mg)	Kd (nmol)	n	B max (fmol/mg)	Kd (nmol)	n
Heart	1590 ± 390	0.92 ± 0.4	5	1260 ± 370	1.57 ± 0.8	7	2 580 ± 1890	1.62 ± 0.9	7
Aorta	24 500 ± 4100	1.9 ± 1.0	9	16 670 ± 3800 **	1.48 ± 0.5	7	20 800 ± 6850	2.68 ± 1.44	7

Table 4. Average B_{max} values **fmoles / mg** protein and K_d values (nM) of [^{3}H]PK 11195 binding to TSPO in aorta and heart homogenates of WT (Bb-Control) and apoE KO mice, fed with standard feed, and feed supplemented with 1% and 3% cholesterol (Chol). One-way analysis of variance ANOVA was used, with Mann-Whitney as the post-hoc, non-parametric test. Data are expressed as mean ± SD; * = p < 0.05, ** = p < 0.01, *** = p < 0.001 vs. control.

3. Discussion

There is strong evidence that accumulation of plasma derived lipoproteins in the arterial wall launches specific cell reactions that account for atherosclerosis process: enhanced NO production, amplification of the inflammatory response, apoptosis, endothelial function impairment, enhanced smooth muscle cell migration and proliferation, and macrophage foam cell formation (Steinberg et al., 2002; Whitman, 2004; Zhao et al., 2008; Singh et al., 2009). Mice lacking apoE have a substantial delay in the metabolism of lipoproteins, particularly VLDL, even fed with a regular standard chow feed (Hoen et al., 2003; Kato et al., 2009). Lesions in apoE-deficient mouse have many features in common with human atherosclerosis, even that the progression can be advantageous in many experimental situations (Dansky et al, 1999). At 26 weeks, atherosclerotic lesions are in the early stages of development, characterized by lipoprotein accumulation, leukocyte gathering, and foam cell formation. This model develops atherosclerotic lesions which progress to occlusion of coronary artery by 8th to 11 months after regular feeding (Piedrahita et al., 1992; Whitman, 2004). Aged (42-54 weeks) apoE KO mice develop intraplaque hemorrhage and plaque instability features, accelerated by feeding westernized diets (Seo et al., 2005; Singh et al., 2009). We found, similar to previous observations, advanced fibrous plaque development accompanying prolonged cholesterol feeding (Figure 1C) in apoE mice but not in WT mice. Another study by Molnar et al. (2005) showed that although high fat feeding induced endothelial cell dysfunction in WT mice, it did not enhance neointimal formation in WT mice. Also in WT rats, a high fat, high cholesterol diet does not appear to lead to atherosclerosis, although modest morphological alterations in the aortic wall could be observed (Dimitrova-Shumkovska et al., 2010a)

We also checked in blood plasma of apoE KO and WT mice the levels of total cholesterol, including triglycerides, high-density lipoprotein and low-density lipoprotein, since it can increase the risk of heart disease and atherosclerosis (Steinberg, 2002; Stocker and Keany, 2004, 2005). Mice naturally have high levels of HDL and low levels of LDL, lacking the cholesterol ester transfer protein, an enzyme responsible for trafficking cholesterol from HDL to VLDL and LDL. As reported also by others previously, we found clear cut differences in abundance of cholesterol related particles between apoE KO mice and WT mice (Table 1), (Hoen et al., 2003; Kato et al., 2009). In particular, each group of apoE KO mice had five times more plasma cholesterol than their WT counterparts. The apoE KO mice also always had higher TAG levels. HDL levels in apoE KO mice supplied with standard feed and 1% cholesterol supplemented diet was also twice as high than in WT mice. Interestingly, 3% cholesterol supplemented diet resulted in a reversal, meaning that HDL levels (i.e. "good" HDL-lipoproteins) in WT mice became twice as high as in apoE KO mice (Table 1). The generally low LDL cholesterol levels in WT mice even with cholesterol supplemented diet may be due to the capability of WT mice to efficiently suppress the percentage of dietary cholesterol absorption by increasing the excretion of gallbladder biliary cholesterol concentration (Sehayek et al., 2000).

We used this model, of apoE KO mice fed with cholesterol supplemented diet that shows well developed atherosclerosis, to assess oxidative stress in the aorta in correlation with TSPO binding density and atherosclerosis. For this purpose, homogenates of the aorta were used for ROS analysis and antioxidant enzymes activities. As accumulation of proatherogenic lipid affects all cell types present within the vascular wall, the response of the entire tissue vs. isolated cells to the hyperlipidemic conditions is relevant as an indication of vascular defense as a whole. The increase in plasma cholesterol levels was paralleled by changes in oxidative stress parameters in WT mice and ApoE KO mice, as discussed in detail below.

An indicator of cellular defence capacity against oxidative stress is the presence of reduced GSH, which we determined in the aorta homogenates after application of feed with cholesterol supplements. As seen in table 3, a reduction of GSH content in was evident compared to the corresponding controls, when 3% cholesterol diet was administered to WT as well as apoE mice. This shows that cholesterol diet regime indeed constitutes an elevated risk factor for ROS formation, due to a reduction in GSH levels in this model. It has been reported that ROS induce vascular cells to express cell adhesion molecules that trigger adhesion of leukocytes to the endothelium, which is part of the initiation atherosclerosis (Yang et al., 2009). Interestingly, it was also found that TSPO expression correlates positively with expression of adhesion molecules (Bode et al., 2012; Veenman et al., 2012). This may suggest that the reduction in TSPO levels seen in this study may counteract adhesion of leukocytes to the endothelium, and thereby prevent initiation atherosclerosis in particular in WT mice.

In accord with the observations of Hoen et al. (2003) that the mRNA levels of many antioxidant enzymes in apoE KO mice are higher (1.5 -5 fold) in the age of 6-15 weeks, compared to aged-matched wild type mice, we also saw that SOD activity were higher in aorta homogenates of apoE mice than those in age-matched WT mice (Table 3). Their hypothesis is that the aorta compensates for the oxidative stress induced by atherogenic stimuli, by stimulating the expression of antioxidant enzymes, thereby delaying the process of atheroma plaque formation. The latter was supported by Yang et al. (2004, 2009) providing evidence that over expression of catalase and superoxide dismutase delayed the development of atherosclerosis in apoE KO mice.

To determine the potential involvement of the TSPO in effects of apoE dysregulation, we studied TSPO binding density in heart and aorta of apoE KO mice (B6.129P2-apoE*tm1* N11) versus their wild type (WT) background mice, with and without inclusion of 1% and 3% cholesterol to the diet. TSPO has been detected in heart of normal mice before, and we found comparable levels in our control animals (Hashimoto et al., 1989; Weizman et al., 1992; Fares et al., 1990; Katz et al., 1994; Dumont et al., 1999). To our knowledge the present study is the first study regarding TSPO binding density in the aorta of mice, which are quite high (even comparable to TSPO levels in adrenal of rats (Gavish and Fares, 1985; Gavish et al., 1999). We found that enhanced cholesterol levels in the diet can result in reduced TSPO binding density in the aorta and heart of WT mice, as well as in the aorta of apoE mice (Table 4). The present study indicates that there is negative correlation between ROS parameters in heart tissue and TSPO binding density in cholesterol fed WT mice. Namely, in the heart of WT mice, the "steady state" levels of lipid peroxides (TBARs) showed a 2.5 fold enhancement after 3% cholesterol supplemented diet vs. a 1/3 fold enhancement in the group with 1% cholesterol supplemented diet. Regarding oxidized proteins in the heart tissues of WT mice fed with cholesterol supplements, AOPP and proteins carbonyls showed increases of 40% and 35%, respectively, regardless of the cholesterol percentage (data not shown). Such a relation between ROS parameters and TSPO binding density is not apparent in apoE mice, since in apoE mice little effect is seen on TSPO binding density.

Also in a previous study, enhanced plasma lipid levels due to HFHC diet supplied to rats, enhanced oxidative stress parameters and decreased indicators for antioxidant activity in the aorta, which were associated with reduced TSPO density in this organ (Dimitrova-Shumkovska et al., 2010a). Notably, wild type rats are not prone to develop atherosclerosis even when subjected to HFHC diet (Dimitrova-Shumkovska et al., 2010a). We have shown that reduction of TSPO expression by genetic manipulation in vitro in cell culture reduces mitochondrial ROS generation (Veenman et al., 2008, 2012; Zeno et al., 2009). We have discussed previously that the reduced TSPO levels accompanying atherogenic challenges may be a compensatory mechanism to counteract oxidative stress in the aorta and liver (Dimitrova-Shumkovska et al., 2010 a, b, c). This would be in effect similar to increased levels of SOD observed, which also counteract oxidative stress (see above). Our present study suggests that reduced TSPO binding density as observed in WT mice subjected to cholesterol supplemented diet may counteract oxidative stress as one mechanism to attenuate the development of atherosclerosis. As TSPO binding density is not affected in apoE mice subjected to cholesterol supplemented diet mentioned TSPO dependent mechanism is not available for apoE KO mice to counteract development of atherosclerosis. Presently, it is not known which components of the vascular wall, i.e. mast cells, smooth muscular, or dermal vascular endothelial cells, would be important for the potential correlation between TSPO expression, oxidative stress, and atherosclerosis (Stoebner et al., 1999; 2001; Morgan et al., 2004; Veenman and Gavish, 2006; Dimitrova-Shumkovska et al., 2010 a, b, c).

It can be assumed from the present study, that oxidative stress parameters do not absolutely correlate with the development of atherosclerotic lesions (because supplementation with 1% of cholesterol to the diet does not affect oxidative stress), but the absolute levels of cholesterol do correlate with atherosclerotic development. Nonetheless, enhancement of cholesterol percentage from 1% to 3% in the diet resulted in significant increases in ROS parameters of WT and apoE KO mice in comparison to their control groups, and also provoked advanced lesion formation in aortic intimae in apoE KO mice fed a 3% cholesterol supplemented diet (but not in WT mice). TSPO binding density is reduced due to cholesterol intake in particular in WT mice and such changes in TSPO binding density in WT mice are in negative correlation with oxidative stress measured in heart and aorta. We believe the reductions in TSPO binding density in WT mice are compensatory for oxidative stress and atherosclerotic development. Thus, the lack of a significant decrease in TSPO binding density in the aorta of 3% cholesterol fed apoE KO mice may actually correlate with the enhanced atherosclerosis in this model. The capability of apoE KO mice fed with 1% cholesterol to reduce TSPO binding density in the aorta may present a rudimentary anti-atherosclerosis protective capacity. In conclusion, this study is in accord with previous studies suggesting that reductions in arterial TSPO binding density are part of a mechanism counteracting the development of atherosclerosis. A question is how the presence of apoE, in combination with enhanced dietary cholesterol levels, can result in suppression of TSPO binding density. It is also important to find out how in a mechanistic sense a reduction in TSPO levels can contribute to self protection against the development of atherosclerosis.

Explanation of abbreviations and symbols: ACS, acute coronary syndrome; ANOVA, analysis of variance; (AOPPs), advanced oxidation protein products; ApoE-/- KO, apolipoprotein E knockout mice; cAMP, adenosine 3,5-cyclic monophosphate; CBR, central-type benzodiazepine receptor; DBI, Diazepam Binding Inhibitor; CAM, cell adhesion molecule; CVD, cardiovascular disease; HDL, high-density lipoprotein; HFHC- high fat high cholesterol diet; HMGCoA, 3-hydroxy-3-methylglutaryl coenzyme A reductase; H_2O_2, hydrogen peroxide; Hb, hemoglobin; IL-1, interleukin-1 (IL-2, etc.); kDa, kilodalton; Kd, equilibrium dissociation constant; Km, equilibrium constant related to Michaelis-Menten kinetics (similarly, Kd, Ka, Keq, Ks); LDL, low density lipoproteins; mPTP, mitochondrial permeability transition pore; MCP-1, monocyte chemoatractant proteins-1; NADP, nicotinamide adenine dinucleotide phosphate; NADH, reduced nicotinamide adenine dinucleotide; PBR, peripheral-type benzodiazepine receptor; PC protein carbonyls; PK 11195, 1-(2- chlorophenyl)-N-methyl-N-(1-methyl-prop1)-3 isoquinolinecarboxamide; ONOO-, peroxinitrite ; Ro5-4864, (4'- chlorodiazepam); ROS , reactive oxygen species; SOD, superoxide dismutase activity; TBARs, thiobarbituric acid reactive substances; TNF, tumor necrosis factor; TSPO, 18 kDa translocator protein; VCAM-1, vascular cell adhesion molecule; VSMCs, vascular smooth muscle cells.

Author details

Jasmina Dimitrova-Shumkovska*
Institute of Biology, Department of Experimental Biochemistry and Physiology, Faculty of Natural Sciences and Mathematics, Ss. Cyril and Methodius University - Skopje, Republic of Macedonia

Leo Veenman, Inbar Roim and Moshe Gavish
Rappaport Family Institute for Research in the Medical Sciences, Technion-Israel Institute of Technology, Department of Molecular Pharmacology, Haifa, Israel

4. References

Anholt, R.R., De Souza, E.B., Kuhar, M.J., & Snyder, S.H. (1985). Depletion of peripheral-type benzodiazepine receptors after hypophysectomy in rat adrenal gland and testis. *Eur J Pharmacol.* 110:41-6.

Aono, M., Bennett, E.R., Kim, K.S., Lynch, J.R., Myers, J., Pearlstein, R.D., Warner, D.S., & Laskowitz, D.T. (2003). Protective effect of apolipoprotein E-mimetic peptides on N-methyl-D-aspartate excitotoxicity in primary rat neuronal-glial cell cultures. *Neuroscience.* 116:437-45.

Banati, R.B., Myers, R., & Kreutzberg, G.W. (1997). PK ('peripheral benzodiazepine')--binding sites in the CNS indicate early and discrete brain lesions: microautoradiographic detection of [3H] PK11195 binding to activated microglia. *J Neurocytol.* 26:77-82.

Beffert, U., Stolt, P.C., & Herz, J. (2004). Functions of lipoprotein receptors in neurons. *J Lipid Res.* 45:403-9.

Benavides, J., Malgouris, C., Imbault, F., Begassat, F., Uzan, A., Renault, C., Dubroeucq, M.C., Gueremy, C., & Le Fur, G. (1983). "Peripheral type" benzodiazepine binding sites in rat adrenals: binding studies with [3H] PK 11195 and autoradiographic localization. *Arch Int Pharmacodyn Ther.* 266:38-49.

Bird, J.L., Izquierdo-Garcia, D., Davies, J.R., Rudd, J.H., Probst, K.C., Figg, N, Clark, J.C., Weissberg, P.L., Davenport, A.P., & Warburton, E.A. (2010). Evaluation of translocator

---

* Jasmina Dimitrova-Shumkovska and Leo Veenman contributed equally to this book chapter.

protein quantification as a tool for characterising macrophage burden in human carotid atherosclerosis. *Atherosclerosis*. 210(2):388-91.

Blue, M.L., Williams, D.L., Zucker, S., Khan, S.A., & Blum, C.B. (1983). Apolipoprotein E synthesis in human kidney, adrenal gland, and liver. *Proc Natl Acad Sci U S A*. 80:283-7.

Bode, J., Veenman, L., Caballero, B., Lakomek, M., Kugler, W., Gavish, M. (2012). The 18 kDa translocator protein influences angiogenesis, as well as aggressiveness, adhesion, migration, and proliferation of glioblastoma cells. *Pharmacogenet Genomics*. 22:538-50.

Bono, F., Lamarche, I., Prabonnaud, V., Le Fur, G., & Herbert, J.M. (1999). Peripheral benzodiazepine receptor agonists exhibit potent antiapoptotic activities. *Biochem Biophys Res Commun*. 265:457-61.

Boyles, J.K., Pitas, R.E., Wilson, E., Mahley, R.W., & Taylor, J.M. (1985). Apolipoprotein E associated with astrocytic glia of the central nervous system and with nonmyelinating glia of the peripheral nervous system. *J Clin Invest*. 76:1501-13.

Bribes, E., Bourrie, B., & Casellas, P. (2003). Ligands of the peripheral benzodiazepine receptor have therapeutic effects in pneumopathies in vivo. *Immunol Lett*. 88:241-7.

Buttini, M., Orth, M., Bellosta, S., Akeefe, H., Pitas, R.E., Wyss-Coray, T., Mucke, L., & Mahley, R.W. (1999). Expression of human apolipoprotein E3 or E4 in the brains of Apoe-/- mice: isoform-specific effects on neurodegeneration. *J Neurosci*. 19:4867-80.

Caliguiri, G., Nicoletti, A., Zhou, X., Tornberg, I. &Hansson, G.K. (1999). Effects of sex and age on atherosclerosis and autoimmunity in apoE-deficient mice. *Atheroscler*. 145(2). 301-8.

Camici PG, Rimoldi OE, Gaemperli

O, Libby P. (2012). Non-invasive anatomic and functional imaging of vascular inflammation and unstable plaque. *Eur Heart J*. 33(11):1309-17.

Cedazo-Minquez, A. & Cowburn, R.F. (2001). Apolipoprotein E: a major piece in the Alzheimer's disease puzzle. *J CellMolMed*. 5(3): 254-66.

Chan, L. (1989). The apolipoprotein multigene family: structure, expression, evolution, and molecular genetics. *Klin Wochenschr*. 67:225-37.

Cheng, C., Tempel, D., Van Haperen, R., Van Damme, L., Algur, M., Krams, R., & De Crom, R. (2009). Activation of MMP8 and MMP13 by angiotensin II correlates to severe intra-plaque hemorrages and collagen breakdown in atherosclerotic lesions with a vulnerable phenotype. *Atheroclerosis*. 204, 26-33.

Chia, M.C. (1998). The role of adhesion molecules in atherosclerosis. *Crit Rev Clin Lab Sci*. 35:573-602.

Costopoulos, C., Liew, T.V., & Bennet, M. (2008). Ageing and Atherosclerosis: Mechanisms and Therapeutic Options. *Biochem Pharmacol*. 75, 1251- 1261.

Csont, T., Bereczki, E., Bencsik, P., Fodor, G., Gorbe, A., Zvara, A., Csonka, C., Puskas, L.G., Santha, M., & Ferdinandy, P. (2007). Hypercholesterolemia inceases myocardial oxidative and nitrosative stress thereby leading to cardiac dysfunction in apoB-100 transgenic mice. *Cardiovasc Res*. 76, 100-109.

Cullen, P., Athanasopoulos, T., Dickson, G., & Owen, J.S. (2001). Cell-derived apolipoprotein E (ApoE) particles inhibit vascular cell adhesion molecule-1 (VCAM-1) expression in human endothelial cells. *J Biol Chem*. 276:46011-6.

Dansky, H., Sherri, A., Charlton, S.A., Sikes, J., Heath, S., Simantov, R., Levin, L., Shu, P., Moore,K., Breslow, J., Smith, J.(1999). Genetic background determines the extent of atherosclerosis in ApoE deficient mice. *Arterioscler.Thromb.Vasc.Biol.* 19:1960-1968.

Davignon, J., Gregg, R.E., & Sing, C.F. (1988). Apolipoprotein E polymorphism and atherosclerosis. *Arteriosclerosis.* 8:1-21. 1988.

Davignon, J., Cohn, J.S., Mabile, L., & Bernier, L. (1999). Apolipoprotein E and atherosclerosis:insight from animal to human studies. 286:115-143.

Dimitrova-Shumkovska, J., Veenman, L., Ristoski, T., Leschiner, S., & Gavish, M. (2010a). Chronic High Fat, High Cholesterol Supplementation Decreases 18 kDa Translocator Protein Binding Capacity in Association with Increased Oxidative Stress in Rat Liver and Aorta. *Food and Chem. Toxicol.* 48, 910-921.

Dimitrova-Shumkovska, J., Veenman, L., Ristoski, T., Leschiner, S., & Gavish, M. (2010b). Dimethylbenz [alpha] anthracene induces oxidative stress and reduces the binding capacity of the mitochondrial 18-kDa translocator protein in rat aorta. *Drug Chem. Toxicol.* 33, 337-47.

Dimitrova-Shumkovska, J., Veenman, L., Ristoski, T., Leschiner, S., & Gavish, M. (2010c). Decreases in Binding Capacity of the Mitochondrial 18 kDa Translocator Protein Accompany Oxidative Stress and Pathological Signs in Rat Liver after DMBA Exposure. *Toxicol Pathol.*, 38:957-68.

Dimitrova-Shumkovska, J., Veenman, & Gavish, M. (2012). The 18kDa Translocator Protein as a Potential Participant in Atherosclerosis, in the book : „Atherogenesis", edited by Parthasarathy, S., INTECH, ISBN 978-953-307-992-9

Do Carmo, S., Seguin, D., Milne, R. & Rassart, E. (2002). Modulation of apolipoprotein D and apolipoprotein E mRNA expression by growth arrest and identification of key elements in the promoter. *J Biol Chem.* 277: 5514-23

Dumont , F., De Vos, F., Versijpt, J., Jansen, H.M., Korf, J., Dierckx, R.A., & Slegers, G. (1999). In vivo evaluation in mice and metabolism in blood of human volunteers of [123I]iodo-PK11195: a possible single-photon emission tomography tracer for visualization of inflammation. *Eur J Nucl Med.* 26:194-200.

Dupont-Wallois, L., Soulié, C., Sergeant, N., Wavrant-de Wrieze, N., Chartier-Harlin, M.C., Delacourte, A.,& Caillet-Boudin, M.L. (1997). ApoE synthesis in human neuroblastoma cells. *Neurobiol Dis.* 4:356-64.

Elliot, A.J., Maier, M.A., Moller, A.C., Friedman, R. & Meinhardt, J. (2007). Color and psychological functioning: the effect of red on performance attainment. *J Exp Psychol Gen.* 136(1): 154-68.

Elliott, D.A., Kim, W.S., Jans, D.A., & Garner, B. (2007). Apoptosis induces neuronal apolipoprotein-E synthesis and localization in apoptotic bodies. *Neurosci Lett.* 416:206-10.

Elshourbagy N.A., Liao W.S., Mahley R.W., & Taylor J.M. (1985). Apolipoprotein E mRNA is abundant in the brain and adrenals, as well as in the liver, and is present in other peripheral tissues of rats and marmosets. *Proc Natl Acad Sci* U S A. 82:203-7.

Epstein, S.E., Stabile, E., Kinnaird, T., Lee, C.W., Clavijo, L. & Burnett, M.S. (2004). Janus phenomenon: the interrelated tradeoffs inherent in therapies designed to enhance collateral formation and those designed to inhibit atherogenesis. *Circulation*. 109(23): 2826-31.

Falchi AM, Battetta B, Sanna F, Piludu M, Sogos V, Serra M, Melis M, Putzolu M, & Diaz, G. (2007). Intracellular cholesterol changes induced by translocator protein (18 kDa) TSPO/PBR ligands. *Neuropharmacology*. 53:318-29.

Fares, F., Weizman, A., Pick, C.G. & Gavish, M. (1990). Effect of prenatal and neonatal chronic exposure to phenobarbital on central and peripheral benzodiazepine receptors. *Brain Res*. 506(1): 115-9.

Finsen, B. (2006). Up-regulation of PK11195 binding in areas of axonal degeneration coincides with early microglial activation in mouse brain. *Eur J Neurosci*. 24:991-1000.

Friedewald WT, Levy RI, & Fredrickson DS. (1972). Estimation of the concentration of low-density lipoprotein cholesterol in plasma, without use of the preparative centrifuge. *Clin Chem*; 18:499—500.

Gaemperli, O., Shalhoub, J., Owen, D. R.J., Lamare, F., Johansson, S., Fouladi, N., Davies, A.H., Rimoldi, O.E., Camici, P.G. (2011). Imaging intraplaque inflammation in carotid atherosclerosis with 11C-PK11195 positron emission tomography/computed tomography. *Eur Heart J*. 33(15):1902-1910.

Gaemperli, O., Shalhoub,J., Owen, D.R.J, Lamare,D., Johansson, S., Fouladi, N., Davies, A., Ornella, E., Rimoldi, O.E.,& Camici, P., (2012). Imaging intraplaque inflammation in carotid atherosclerosis with 11C-PK11195 positron emission tomography/computed tomography. *Eur Heart J*. 33(15): 1902-1910.

Gavish, M. & Fares, F. (1985).Solubilization of peripheral benzodiazepine-binding sites from rat kidney. *J Neurosci*. 5:2889-93.

Gavish, M., Bar-Ami, S., & Weizman, R. (1992). The endocrine system and mitochondrial benzodiazepine receptors. *Mol Cell Endocrinol*. 88, 1-13.

Gavish, M. & Weizman, R. (1997). Role of peripheral-type benzodiazepine receptors in steroidogenesis. *Clin Neuropharmacol*. 20:473-81.

Gavish, M., Bachman, I., Shoukrun, R., Katz, Y., Veenman, L., Weisinger, G. & Weizman, A. (1999). Enigma of the peripheral benzodiazepine receptor. *Pharmacol. Review*. 51, 630-646.

Gunten, A.V., Ebbing, K., Imhof, A., Giannakopoulos, P., Kovari, E. (2010). Brain aging in the Oldest-Old. *Curr. Gerontol Geriatr. Res*. 358531.

Han, S.H., Einstein, G., Weisgraber, K.H., Strittmatter, W.J., Saunders, A.M., Pericak-Vance, M., Roses, A.D., & Schmechel, D.E. (1994). Apolipoprotein E is localized to the cytoplasm of human cortical neurons: a light and electron microscopic study. *J Neuropathol Exp Neurol*. 53:535-44.

Harris, F.M., Tesseur, I., Brecht, W.J., Xu, Q., Mullendorff, K., Chang, S., Wyss-Coray, T., Mahley, R.W., & Huang, Y. (2004). Astroglial regulation of apolipoprotein E expression in neuronal cells. Implications for Alzheimer's disease. *J Biol Chem*. 279:3862-8.

Hashimoto, K., Inoue, O., Suzuki, K., Yamasaki, T., & Kojima, M. (1989). Synthesis and evaluation of 11C-PK 11195 for in vivo study of peripheral-type benzodiazepine receptors using positron emission tomography. *Ann Nucl Med*. 3:63-71.

Hasty, A.H., Linton, M.F., Brandt, S.J., Babaev, V.R., Gleaves, L.A., & Fazio, S. (1999). Retroviral gene therapy in ApoE-deficient mice: ApoE expression in the artery wall reduces early foam cell lesion formation. *Circulation.* 99:2571-6.

Heumann, R., Lindholm, D., Bandtlow, C., Meyer, M., Radeke, M.J., Misko, T.P., Shooter, E.,

Hoen, P.A., Van der Lans, C.A., Van Eck, M., Bijsterbosch, M.K., Van Berkel, T.J. & Twisk, J. (2003). Aorta of ApoE-deficient mice responds to atherogenic stimuli by a prelesional increase and subsequent decrease in the expression of antioxidant enzymes. *Circ Res.* 93(3): 262-9.

Hoen, P.A.C., Van der Lans, C.A.C., Van Eck, M., Bijsterbosch, M.K., Van Berkel & T.J.C., Twisk, J. (2003). Aorta of ApoE – deficient mice responds to atherogenic stimuli by a prelesional increase and subsequent decrease in the expression of antioxidant enzymes. *Circ Res.* 93, 262-269.

Hollingsworth, E.B., McNeal, E.T., Burton, J.L., Williams, R.J., Daly, J.W., & Creveling, C.R. (1985). Biochemical characterization of a filtered synaptoneurosome preparation from guinea pig cerebral cortex: cyclic adenosine 3':5'-monophosphate-generating systems, receptors, and enzymes. *J Neurosci.* 5:2240-53.

Insull, W.Jr. (2009). The pathology of atherosclerosis: Plaque development and plaque responses to medical treatment. *Am J Med.* 122:S3-S14.

Johnson, J., Carson,K., Williams, H., Karanam, S., Newby, A., & Angelini, G. (2005). Plaque rupture after short periods of fat feeding in the apolipoprotein E-knockout mouse:model characterization and effects of pravastatin treatment. *Circulation.*111:1422-1430.

Kashyap, V.S., Santamarina-Fojo, S., Brown, D.R., Parrott, C.L., Applebaum-Bowden, D., Meyn, S., Talley, G., Paigen, B., Maeda, N., & Brewer, H.B. Jr. (1995). Apolipoprotein E deficiency in mice: gene replacement and prevention of atherosclerosis using adenovirus vectors. *J Clin Invest.* 96:1612-20.

Kato R, Mori C, Kitazato K, Arata S, Obama T, Mori M, Takahashi K, Aiuchi T, & Takano T. (2009). Transient increase in plasma oxidized LDL during the progression of atherosclerosis in apolipoprotein E knockout mice. *Arterioscler Thromb Vasc Biol.* 29: 33–9.

Klegeris, A., McGeer, E.G., & McGeer, P.L. (2000). Inhibitory action of 1-(2-chlorophenyl)-N-methyl-N-(1-methylpropyl)-3-isoquinolinecarboxam ide (PK 11195) on some mononuclear phagocyte functions. *Biochem Pharmacol.* 59:1305-14.

Kmieć Z. (2001). Cooperation of liver cells in health and disease. *Adv Anat Embryol Cell Biol.* 161:1-151.

Knight-Lozano, C.A., Young, C.G., Burow, D.L., Hu, Z.Y., Uyeminami, D., Pinkerton, K.E., Ischiropulos, H., & Ballinger, S.W. (2002). Cigarette smoke exposure and hypercholesterolemia increase mitochondrial damage in cardiovascular tissues. *Circ Res.* 105, 849-854.

Ko, K.W.S., Paul, A., Ma, K., Li, L., Chan, L. (2005) Endothelial lipase modulates plasma lipoprotein profiles but has no effect on the development od atherosclerosis in ApoE$^{-/-}$ and LDLR$^{-/-}$ mice. *J Lipid Res.* 46:2586-2594.

Krueger, K.E & Papadopoulos, V. (1990). Peripheral-type benzodiazepine receptors mediate translocation of cholesterol from outer to inner mitochondrial membranes in adrenocortical cells. *J Biol Chem.* 265:15015-22.

Krul, F.S., Tikkanen, M.J., & Schonfeld, G. (1988). Heterogeneity of apolipoprotein E epitope expression on human lipoproteins:importance for apolipoprotein E function. *J Lipid Res.* 29:1309-25.

Kunitomo, M., Yamaguchi, Y., Kagota, S., Yoshikawa, N., Nakamura, K., & Shinozuka, K. (2009). Biochemical Evidence of Atherosclerosis Progression Mediated by Increased Oxidative Stress in Apolipoprotein E-Deficient Spontaneously Hyperlipidemic Mice Exposed to Chronic Cigarette Smoke. *J Pharmacol. Sci.* 110, 354-361

Lacapère, J.J. & Papadopoulos, V. (2003). Peripheral-type benzodiazepine receptor: structure and function of a cholesterol-binding protein in steroid and bile acid biosynthesis. *Steroids.* 68:569-85.

Laskowitz, D. T. (2003). APOE genotype and an ApoE-mimetic peptide modify the systemic and central nervous system inflammatory response. *J Biol Chem.* 278:48529-33.

Law, G.L., McGuinness, M.P., Linder, C.C., & Griswold, M.D. (1997). Expression of apolipoprotein E mRNA in the epithelium and interstitium of the testis and the epididymis. *J Androl.* 18:32-42.

Lenfant, M., Zavala, F., Haumont, J., & Potier, P. (1985). [Presence of a peripheral type benzodiazepine binding site on the macrophage; its possible role in immunomodulation] *C R Acad Sci* III. 300:309-14.

Levine, R. L., Garland, D., Oliver, C.N., Amici, A., Climent, I., Lenz, A.G., Ahn, B.W., Shaltiel, S., & Stadman, E.R. (1990). Determination of carbonyl content in oxidatively modified proteins. *Methods in Enzymol.* 186, 464-478.

Libby, P., Ridker, P.M. & Hansson, G.K. (2011). Progress and challenges in translating the biology of atherosclerosis. Review. *Nature* 473:317-325

Linton, MF. & Fazio, S. (1999). Macrophages, lipoprotein metabolism, and atherosclerosis: insights from murine bone marrow transplantation studies. *Curr Opin Lipidol.* 10:97-105.

Liu S.X., Hou, F.F., Guo, Z. J., Nagai, R., Zhang, W.R., Liu, Z.Q., Zhou, Z.M., Di Xie, Wang, G.B., & Zhang, X. (2006). Advanced Oxidation Protein Products Accelerate Atherosclerosis Through Promoting Oxidative Stress and Inflammation. *Arterioscler Thromb Vasc Biol.* 26, 1156-1162.

Lynch, J.R., Tang, W, Wang, H., Vitek, M.P., Bennett ,E.R., Sullivan, P.M., Warner, D.S.,

Mahley, R.W. (1988). Apolipoprotein E: cholesterol transport protein with expanding role in cell biology. *Science.* 240(4582): 622-30.

Mahley, R.W., & Huang, Y. (1999a). Apolipoprotein E: from atherosclerosis to Alzheimer's disease and beyond. *Curr Opin Lipidol.* 10:207-17.

Mahley, R.W. & Ji, Z.S. (1999b). Remnant lipoprotein metabolism: key pathways involving cell-surface heparan sulfate proteoglycans and apolipoprotein E. *J Lipid Res.* 40:1-16.

Mahley, R.W. & Rall, S.C. Jr. (2000). Apolipoprotein E: far more than a lipid transport protein. *Annu Rev Genomics Hum Genet.* 1:507-37.

Mahley, R.W., Huang, Y., & Weisgraber, K.H. (2006).Putting cholesterol in its place: apoE and reverse cholesterol transport. *J Clin Invest.* 116:1226-9.

Marchetti, P., Trincavelli, L., Giannarelli, R., Giusti, L., Coppelli, A., Martini, C., & Marshall, J.S. (2004). Mast-cell responses to pathogens. *Nat Rev Immunol*. 4:787-99.

Mazurika, C., Veenman, L., Weizman, R., Bidder, M., Leschiner ,S., Golani, I., Spanier, I., Weisinger, G., & Gavish, M. (2009). Estradiol modulates uterine 18 kDa translocator protein gene expression in uterus and kidney of rats. *Mol Cell Endocrinol*. 307, 43-49.

McAdoo, J.D., Warner, D.S., Goldberg, R.N., Vitek, M.P., Pearlstein, R., & Laskowitz, D.T. (2005). Intrathecal administration of a novel apoE-derived therapeutic peptide improves outcome following perinatal hypoxic-ischemic injury. *Neurosci Lett*. 381:305-8.

Milner, P., Bodin, P., Guiducci, S., Del Rosso, A., Kahaleh, M.B., Matucci-Cerinic, M., & Burnstock, G. (2004). Regulation of substance P mRNA expression in human dermal microvascular endothelial cells. *Clin Exp Rheumatol*. 22:S24-7.

Morgan, J., Oseroff, A.R., & Cheney, R.T. (2004). Expression of the peripheral benzodiazepine receptor is decreased in skin cancers in comparison with normal skin. *Br J Dermatol*. 151, 846-56.

Nakashima, Y., Raines, E.W., Plump, A.S., Breslow, J.L., & Ross, R. (1998). Upregulation of VCAM-1 and ICAM-1 at atherosclerosis-prone sites on the endothelium in the ApoE-deficient mouse. *Arterioscler Thromb Vasc Biol*. 18:842-51.

Navalesi, R. & Lucacchini, A. (1996). Characterization of peripheral benzodiazepine receptors in purified large mammal pancreatic islets. *Biochem Pharmacol*. 51:1437-42.

Noeman, S.A., Hamooda, H.E., & Baalash, A.A. (2011). Biochemical study of oxidative stress markers in the liver, kidney and heart of high fat diet induced obesity in rats. *Diabetol Metab Syndr*. 3(1), 17.

Onyimba, J.A., Coronado, M.J., Garton, A.E., Kim, J.B., Bucek, A., Bedja, D., Gabrielson, K.L., Guilarte, T.R., & Fairweather, D. (2011). The innate immune response to coxsackievirus B3 predicts progression to cardiovascular disease and heart failure in male mice. *Biol Sex Differ*. 21; 2:2.

Papadopoulos, V., Baraldi, M., Guilarte, T.R., Knudsen, T.B., Lacapère, J.J., Lindemann, P., Norenberg, M.D., Nutt, D., Weizman, A., Zhang, M.R., & Gavish, M. (2006). Translocator protein (18kDa). New nomenclature for the peripheral-type benzodiazepine receptor based on its structure and molecular function. *Trends Pharmacol Sci*. 27, 402-9.

Papadopoulos, V., Berkovich, A., Krueger, K.E., Costa, E., & Guidotti, A. (1991). Diazepam binding inhibitor and its processing products stimulate mitochondrial steroid biosynthesis via an interaction with mitochondrial benzodiazepine receptors. *Endocrinology*. 129:1481-8.

Pedersen, M.D., Minuzzi, L., Wirenfeldt ,M., Meldgaard, M., Slidsborg, C., Cumming, P., Potter, H., Wisniewski, T. (2012). Apolipoprotein E: Essential catalyst of the Alzheimer amyloid cascade. *Int J Alzheimers Dis*. 489428.

Quinn, C.M., Kågedal, K., Terman, A., Stroikin U., Brunk, U.T., Jessup, W., & Garner, B. (2004). Induction of fibroblast apolipoprotein E expression during apoptosis, starvation-induced growth arrest and mitosis. *Biochem J*. 378:753-61. 2004.

Rosenfeld, M.E., Polinsky, P., Virmani, R., Kauser, K., Rubanyi,G., Schwartz, S.M. (2000). Advanced atherosclerotic lesions in the innominate artery of the ApoE knockout mouse. *Arterioscler. Thromb. Vasc. Biol.*20:2587-2592.

Ross, J.S., Stagliano, N.E., Donovan, M.J., Breitbart, R.E., & Ginsburg, G.S. (2001). Atherosclerosis And Cancer: Common Molecular Pathways Of Disease Development And Progression. *Ann NY Acad Sci.* 947, 271–292Ruff, M.R., Wahl S.M., Mergenhagen, S., & Pert, C.B. (1985). Opiate receptor-mediated chemotaxis of human monocytes. *Neuropeptides.* 5:363-6.

Ross, R & Glomset, J.A. (1973). Atherosclerosis and the arterial smooth muscle cell: Proliferation of smooth muscle is a key event in the genesis of the lesions of atherosclerosis. *Science* 180: 1332-1339.

Ross, R., (1999). Atherosclerosis an Inflammatory Disease. *N Engl J Med.* 340, 115-26.

Ryu, J.K., Choi, H.B., & McLarnon, J.G. (2005). Peripheral benzodiazepine receptor ligand PK11195 reduces microglial activation and neuronal death in quinolinic acid-injected rat striatum. *Neurobiol Dis.* 20:550-61.

Schafer, F.Q., Buettner, G.R. (2001) . Redox environment of the cell as viewed through the redox state of the glutathione disulfide/glutathione couple. *Free Radic Biol Med.* 30:1191-212.

Sehayek, E., Shefer, S., Nguyen, L.B., Ono, J.G., Merkel, M., & Breslow, J.L. (2000). Apolipoprotein E regulates dietary cholesterol absorption and biliary cholesterol excretion: studies in C57BL/6 apolipoprotein E knockout mice. *Proc Natl Acad Sci U S A.* 97:3433-7.

Seo, T., Qi, K., Chang, C., Liu, Y., Worgall, T.S., Ramakrishan, R., & Deckelbaum, R.J. (2005). Saturated fat-rich diet enhances selective uptake of LDL cholesteryl esters in the arterial wall. *J Clin Invest.* 115, 2214-2222.

Shargorodsky, L., Veenman, L., Caballero, B., Pe'er, Y., Leschiner, S., Bode, J., Gavish, M. (2012). The nitric oxide donor sodium nitroprusside requires the 18 kDa Translocator Protein to induce cell death. *Apoptosis.* 17:647-65.

Sheng, H., Laskowitz, D.T., Bennett, E., Schmechel, D.E., Bart, R.D., Saunders, A.M., Pearlstein, R.D., Roses, A.D., & Warner, D.S. (1998). Apolipoprotein E isoform-specific differences in outcome from focal ischemia in transgenic mice. *J Cereb Blood Flow Metab.* 18:361-6.

Singh, V., Tiwari, R.L., Dikshit, M., & Barthwal, M.K. (2009). Models to Study Atherosclerosis: A Mechanistic Insight. *Curr Vasc Pharmacol.* 7, 75-109.

Stannard, A.K., Riddell, D.R., Sacre, S.M., Tagalakis, A.D., Langer, C., von Eckardstein, A., &

Steinberg, D., Parthasarathy, S., Carew, T.E., Khoo, J.C., & Witztum, J.L. (1989). Beyond cholesterol: modifications of low-density lipoprotein that increase its atherogenicity. *N Engl J Med.* 320:915-924

Steinberg, D. (2002). Atherogenesis in perspective: hypercholesterolemia and inflammation as partners in crime. *Nat Med* 8, 1211–1217.

Stephenson, D.T., Schober, D.A., Smalstig, E.B., Mincy, R.E., Gehlert, D.R., & Clemens, J.A. (1995). Peripheral benzodiazepine receptors are colocalized with activated microglia following transient global forebrain ischemia in the rat. *J Neurosci.* 15:5263-74.

Stocker, R & Keaney, JF Jr. (2004). Role of oxidative modifications in atherosclerosis. *Physiol Rev.* 84,1381–478.

Stocker, R & Keaney, J.F. Jr. (2005). New insights on oxidative stress in the artery wall. *J Thromb Haemost* 3, 1825-1834.

Stoebner, P.E., Carayon, P., Penarier, G., Fréchin, N., Barnéon, G., Casellas, P., Cano, J.P., Meynadier, J., & Meunier, L. (1999). The expression of peripheral benzodiazepine receptors in human skin: the relationship with epidermal cell differentiation. *Br J Dermatol.* 140:1010-6.

Stoebner, P.E., Carayon, P., Casellas, P., Portier, M, Lavabre-Bertrand, T., Cuq, P., Cano, J.P., Meynadier, J., & Meunier L. (2001). Transient protection by peripheral benzodiazepine receptors during the early events of ultraviolet light-induced apoptosis. *Cell Death Differ.* 8:747-53.

Strittmatter, W.J. & Bova Hill ,C. (2002). Molecular biology of apolipoprotein E. *Curr Opin Lipidol.* 13:119-23.

Taupin, V., Toulmond, S., Serrano, A., Benavides, J., & Zavala, F. (1993). Increase in IL-6, IL-1 and TNF levels in rat brain following traumatic lesion. Influence of pre- and post-traumatic treatment with Ro5 4864, a peripheral-type (p site) benzodiazepine ligand. *J Neuroimmunol.* 42:177-85.

Thoenen, H. (1987). Differential regulation of mRNA encoding nerve growth factor and its receptor in rat sciatic nerve during development, degeneration, and regeneration: role of macrophages. *Proc Natl Acad Sci* U S A. 84:8735-9.

Torres, S.R., Fröde, T.S., Nardi, G.M., Vita, N., Reeb, R., Ferrara, P., Ribeiro-do-Valle, R.M.,Farges, R.C. (2000). Anti-inflammatory effects of peripheral benzodiazepine receptor ligands in two mouse models of inflammation. *Eur J Pharmacol.* 408:199-211.

Truskey, G.A., Herrmann, R.A., Kait, J., & Barber, K.M. (1999). Focal increases in vascular cell adhesion molecule-1 and intimal macrophages at atherosclerosis-susceptible sites in the rabbit aorta after short-term cholesterol feeding. *Arterioscler Thromb Vasc Biol.* 19:393-401.

Veenman, L. & Gavish, M. (2000). Peripheral-type benzodiazepine receptors: Their implication in Brain Disease. *Drug Dev Res* 50:355-370

Veenman, L. & Gavish, M. (2006). The peripheral-type benzodiazepine receptor and the cardiovascular system. Implications for drug development. *Pharmacol Ther.*110, 503-24.

Veenman, L., Papadopoulos, V., & Gavish, M. (2007). Channel-like functions of the 18-kDa translocator protein (TSPO): regulation of apoptosis and steroidogenesis as part of the host-defence response. *Curr Pharm Des.* 13, 2385-405.

Veenman, L., Shandalov, Y., & Gavish, M. (2008). VDAC activation by the 18 kDa translocator protein (TSPO), implications for apoptosis. *J Bioenerg Biomembr.* 40(3):199-205.

Veenman, L., Gavish, M. (2012). The role of 18 kDa mitochondrial translocator protein (TSPO) in programmed cell death, and effects of steroids on TSPO expression. *Curr Mol Med.* 12(4): 398-412.

Veenman, L., Bode, J., Gaitner, M., Caballero, B., Pe'er, Y., Zeno, S., Kietz, S., Kugler, W., Lakomek, M., Gavish, M. (2012) Effects of 18-kDa translocator protein knockdown on gene expression of glutamate receptors, transporters, and metabolism, and on cell viability affected by glutamate. *Pharmacogenet Genomics.* 22:606-19.

Veiga, S., Azcoitia, I., Garcia-Segura, L.M. (2005). Ro5-4864, a peripheral benzodiazepine receptor ligand, reduces reactive gliosis and protects hippocampal hilar neurons from kainic acid excitotoxicity. *J Neurosci Res.* 80:129-3.

Vowinckel, E., Reutens, D., Becher, B., Verge, G., Evans, A., Owens, T., Antel, J.P.(1997). PK11195 binding to the peripheral benzodiazepine receptor as a marker of microglia activation in multiple sclerosis and experimental autoimmune encephalomyelitis. *J Neurosci Res.* 50:345-53.

Waterfield, J.D., McGeer, E.G., McGeer, P.L. (1999). The peripheral benzodiazepine receptor ligand PK 11195 inhibits arthritis in the MRL-lpr mouse model. Rheumatology (Oxford). 38:1068-73.

Weisgraber, K.H., Roses, A.D., Strittmatter, W.J.(1994). The role of apolipoprotein E in the nervous system. *Curr Opin Lipidol.* 5:110-6.

Whitman, S.C. (2004). A practical approach to using mice in atherosclerosis research. *Clin Biochem Rev.* 25(1), 81-93.

Wilms, H., Claasen, J., Röhl, C., Sievers, J., Deuschl, G., & Lucius, R. (2003). Involvement of benzodiazepine receptors in neuroinflammatory and neurodegenerative diseases: evidence from activated microglial cells in vitro. *Neurobiol Dis.* 14:417-24.

Wojta, J., Huber, K., & Valent, P. (2003). New aspects in thrombotic research: complement induced switch in mast cells from a profibrinolytic to a prothrombotic phenotype. *Pathophysiol Haemost Thromb.* 33:438-41.

Woods, A.A., Linton, S.M. & Davies, M.J. (2003). Detection of HOCl-mediated protein oxidation products in the extracellular matrix of human atherosclerotic plaques. *Biochem.* 370: 729-35.

Xu, P.T., Schmechel, D., Rothrock-Christian, T., Burkhart, D.S., Qiu, H.L., Popko, B., Sullivan, P., Maeda, N, Saunders, A.M., Roses, A.D., & Gilbert, J.R. (1996). Human apolipoprotein E2, E3, and E4 isoform-specific transgenic mice: human-like pattern of glial and neuronal immunoreactivity in central nervous system not observed in wild-type mice. *Neurobiol Dis.* 3:229-45.

Yang, H., Roberts, J.L., Shi, M.J., Zhou, C., Ballard, B.R., Richardson, A .& Guo, Z.M. (2004). Retardation of atherosclerosis by overexpression of catalase or both Cu/Zn-superoxide dismutase and catalase in mice lacking apoliporotein E. *Circ Res.* 95, 1075-1081.

Young, C.G., Knight, C., Vickers, K.C., Westbrook, D., Madamanchi, N.R., Runge, M.S., Ischiropoulos, H., & Ballinger, S.W. (2004). Differential effects of exercise on aortic mitochondria. *Am J Physiol Heart Circ Physiol.* 288, H1683-H1689.

Zeno, S., Zaaroor, M., Leschiner, S., Veenman, L., & Gavish, M. (2009). CoCl(2) induces apoptosis via the 18kDa translocator protein in U118MG human glioblastoma cells. *Biochem.* 48, 4652-61.

Zeno, S., Veenman, L., Katz, Y., Bode, J., Gavish, M., & Zaaroor, M. (2012). The 18 kDa mitochondrial Translocator Protein (TSPO) prevents accumulation of protoporphyrin IX. Involvement of reactive oxygen species. *Curr Molecular Medicine.* 12(4): 494-501.

Zhao, Y., Kuge, Y., Zhao, S., Strauss, W.H., Blankenberg, F.G., & Tamaki, N. (2008). Prolonged high-fat feeding enhances aortic 18F-FDG and 99mTc-annexin A5 uptake in apolipoprotein E-deficient and wild type C57BL/6J mice. *J Nucl Med.* 49, 1707-1714.

Chapter 6

Autophagy Regulates Lipid Droplet Formation and Adipogenesis

Yasuo Uchiyama and Eiki Kominami

Additional information is available at the end of the chapter

1. Introduction

Proteolysis in eukaryotic cells can be separated into two major pathways: one is mediated by the ubiquitin-proteasome system and by the autophagy-lysosome system. Substrates of lysosomes may be taken through heterophagocytosis, endocytosis, or autophagy into heterophagosomes, early endosomes, or autophagosomes, which receive lysosomal enzymes via transporting vesicles from the trans-Golgi network or lysosomes and become heterophagolysosomes, late endosomes, or autolysosomes [1-4]. Different from endocytosis and heterophagocytosis, substrates of autophagy are limited to intracellular constituents and include various membranous organelles together with a part of the cytoplasm [3]. Until recently, 18 autophagy-related proteins (Atgs) have been shown to be involved in autophagosome formation, although more than 30 Atgs have been uncovered to regulate autophagy [2, 5]. Such autophagy contributes to the maintenance of cellular homeostasis [6]. Impairment of autophagy, therefore, causes severe degenerative alterations in various tissue cells [1, 4, 7]. In addition to the maintenance of basal cellular metabolism, autophagy is induced in response to various stresses such as starvation and diseases [8-13].

Since one of the major sources in living organisms is lipid, the metabolism of lipid is finely regulated. Besides adipose tissues, neutral lipids, most of which are mainly triacylglycerol (TAG) and cholesterol ester (CE) in cells, are stored in a sort of inclusion body, called the lipid droplet or lipid body [14-16]. Under nutrient-rich situations, excess fatty acids may be converted to TAG through lipogenesis and stored in lipid droplets, whereas lipid droplets under starvation conditions may be a source to produce lipids by lipolysis for cell usage [16]. Under stress or starvation states, proteins are also degraded to produce an amino acid pool that is used in part for energy metabolism through glyconeogenesis, while free fatty acids (FFAs) from adipocytes are delivered to hepatocytes and cardiac myocytes where they are used as an energy source via β-oxidation [15, 16]. At the initial stage during fasting,

FFAs are also converted to triacylglycerol (TAG) and stored as lipid droplets (LDs) that are used for an energy source when starvation continues.

It has been shown that loss of Atg7 largely suppresses LD formations in hepatocytes and cardiac myocytes 24 hours after the start of starvation, although numerous LDs accumulate in normal hepatocytes and cardiac myocytes under the same conditions [11]. Moreover, a mouse model with a targeted deletion of atg7 in adipose tissue has been generated; the mutant mice were slim and contained only 20% of the mass of white adipose tissue (WAT) found in wild-type mice [17]. These mutant mice exhibit a high sensitivity to insulin that reduces low fed plasma concentrations of FFAs, and also exhibit a marked decrease in plasma concentrations of leptin but not adiponectin, and lower plasma concentrations of TAG. LDs, initially considered inert lipid deposits, have gained the classification of cytosolic organelles during the last decade due to their defined composition and the multiplicity of specific cellular functions in which they are involved [18]. At present, it remains largely unknown how autophagy is involved in LD metabolism, although lipophagy may occur in cells.

One thing that we have found is that a lipidated form of LC3, representing the Atg8 family of proteins, is localized on the surface of LDs and also in LD fractions, in addition to ADRP and perilipin, representing the PAT family of proteins that cover the surface of LDs. In this review we will introduce the LC3 conjugation system that is involved in lipid metabolism via LD formation..

2. Microtubule-associated protein 1A/1B light chain 3 (LC3) localizes not only to autophagosomal membranes but also on the surface membrane of LDs in hepatic and cardiac tissues under starvation conditions

It is well known that hepatocytes are morphologically and functionally different between the periportal and perivenous regions; periportal hepatocytes are glycogenic and lipolytic, while perivenous hepatocytes are glycolytic and lipogenic [19]. Lysosomes in hepatocytes are also more abundant in the perivenous region than in the periportal region [19, 20].

Such a morphological difference is largely altered when mice are starved for 24 or 48 hours; numerous autophagosomes that contain part of the cytoplasm and possess the cisternal or double isolation membranes and autolysosomes appear near bile canaliculi in hepatocytes 24 hours after the commencement of starvation and those that in some cases contain mitochondria are seen 48 hours later [4, 9]. In addition to autophagosomes and autolysosomes, LDs accumulate abundantly in the cytoplasm of hepatocytes. When livers of conditional Atg7-deficient mice (Atg7[Flox/Flox]: Albumin-Cre) at 22 days, or 6 or 8 weeks of age were examined, positive staining for LDs is sparsely detected in the hepatocytes under starvation conditions for 12 or 24 hours, although they are abundant in the cytoplasm of the hepatocytes of the control littermate mouse liver (Fig. 1). Size and amount of stained LDs were much smaller in hepatocytes deficient in Atg7 than in the control hepatocytes. When examined by electron microscopy, the diameter of LDs in the control hepatocytes is various and ranged up to 2.61 μm (the mean diameter (± SD)) is 1.12 ± 0.17 μm). Different from the control, mutant hepatocytes mainly possess smaller LDs whose diameter ranged up to 0.83 μm (the average diameter is 0.19 ± 0.17 μm). Moreover, small LD-like bodies were observed in the luminal space of rough surfaced endoplasmic reticulum (rER), Golgi cisternal ends, and vesicles near the Golgi apparatus of the cells, indicating that the lumenally-sorted LDs are normally produced in the rER of both mutant and control hepatocytes. Importantly, the total TAG amount in Atg7-deficient liver and control livers is half of that in the control liver. These data indicate that the conjugation system of LC3 by Atg7 is required for the formation of LDs.

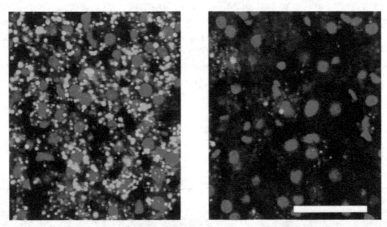

Figure 1. BODIPY staining (green) of hepatocytes obtained from control littermate (left) and Atg7-deficient (right) mice at the age of 6 weeks. Mice were housed under starvation conditions for 24 hours. BODIPY-positive LDs are abundant in hepatocytes from control littermate mouse, while positive LDs largely disappear from the Atg7-deficient hepatocytes. Bar= 50 μm

3. Inhibition of LD formation in hepatocytes deficient in Atg7 after deprivation

It has been shown that autophagy may play an important role in normal adipogenesis and that inhibition of autophagy by disrupting the *atg7* gene has a unique anti-obesity and insulin sensitization effect [17]. LDs are ubiquitous in eukaryotic cells, while excess free fatty acids and glucose in plasma are converted to TAG and stored as LDs (Fig. 1). However, the mechanism for the generation and growth of LDs in cells is largely unknown. As stated above, Atg7 that mediates LC3 lipidation and is essential for autophagy is involved in LD formation [4, 9]. LD formation accompanied by accumulation of TAG induced by starvation is largely suppressed in hepatocytes that cannot execute autophagy.

It is well known that LC3, microtubule-associated protein A/B light chain 3, is localized on the surface of the isolation membrane when starvation is induced [9]. Using GFP-LC3 transgenic mice, GFP-LC3 becomes dot-shaped, cap-shaped and ring-shaped in hepatocytes and cardiac myocytes under starvation conditions [21]. When immunostained for LC3 under starvation conditions, positive staining is, as shown by GFP-LC3 in TG mice, dot-shaped, cap-shaped, and ring-shaped in hepatocytes and cardiac myocytes (Figs. 2 and 3, Tables 1 and 2). Interestingly, positive staining of LC3 in cardiac myocytes is longitudinally arrayed in parallel to myofilaments. In both hepatocytes and cardiac myocytes, LDs are abundant 24 hours after the onset of starvation [9]. In particular, electron microscopic observations show that no clear-cut autophagosomes are detected in cardiac myocytes, although many large LDs are arranged longitudinally in parallel to the array of myofilaments together with mitochondria. This arrangement of LDs in cardiac myocytes is very similar to that of LC3-positive granules. To examine the relationship between staining patterns of LC3 and LDs in both hepatocytes and cardiac myocytes, double staining for perilipin with LC3 was performed. The results indicate that perilipin-positive LDs are also immunopositive for LC3 on the surface of LDs in both hepatocytes and cardiac myocytes [9]. In hepatocytes, there is also dotted staining of LC3 that is free of LD staining, whereas most LD-positive staining is co-localized with LC3 in cardiac myocytes (Figs. 2 and 3, Tables 1 and 2).

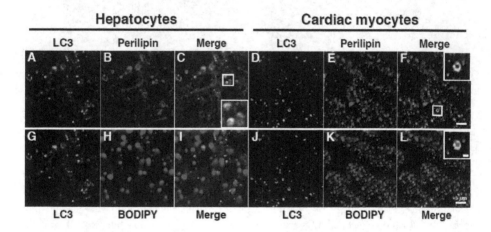

Figure 2. LC3 localizes not only to autophagosomes but also LDs in liver and heart tissues of mice under starvation conditions for 24 hours. (A-L) Double staining of LC3 (green) (A, D, G, J) and perilipin (B, E) or BODIPY (H, K) (red) in liver (A-C, G-I) and heart (D-F, J-L) tissues. Ring-shaped structures that were costained for LC3 and perilipin in boxed areas (C, F) are enlarged in insets. One of the ring-shaped structures that were costained for LC3 and BODIPY in a boxed area (L) was enlarged in an inset. Bar indicates 5 μm (1 μm in inset). This figure is referred from Biochemical and Biophysical Research Communications (Shibata et al., 382 (2009) 419–423).

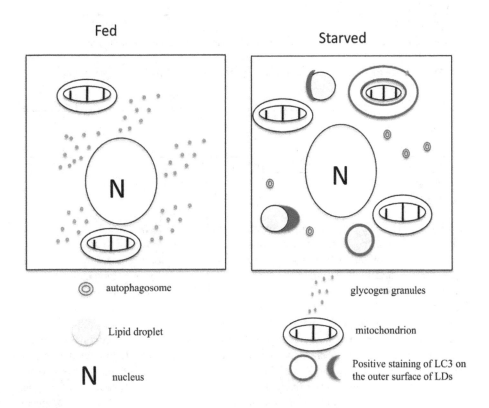

Figure 3. Schematic demonstrations of LC3 staining in cells from mice under fed (left panel) and starved (right panel) conditions. When mice are fed, cells start to produce and store glycogen granules, whereas under starved conditions, lipid droplets are increased and many autophagic granules with double membranes increase in the cytoplasm. In this situation, the lipidated form of LC3 is attached to the isolation membrane of autophagosomes (blue color) and the surface membrane of LDs (yellow). Starvation is further continued, and some mitochondria are enwrapped by the isolation membrane (mitophagy).

124   Lipid Metabolism

Starvation	hepatocytes				cardiac myocytes			
	Apg		LD		Apg		LD	
	24 h	48 h	24 h	48 h	24 h	48 h	24 h	48 h
	++	++	++	++	+/-	+/-	++	+
LC3 (ICH)	++	++	++	++	+/-	-	+	+/-
LC3-II (WB)	++	++	++	++	++	+/-	+	+/-
Mitophagy (EM)	+/-	+			not detected			

Apg, autophagosome; EM, electron microscopy; h, hours; ICH, immunocytochemistry; ++, highly positive; +, positive; +/-, rare; WB, Western blot

Table 1. Morphological changes detected in hepatocytes and cardiac myocytes after starvation

	Liver	heart
lipid fraction		
ADRP	++	++
LC3-II	+	+
Staining of LD in LD fraction from livers with anti-LC3	++	

++, highly positive; +, positive

Table 2. Proteins detected in lipid fractions

In fact, cytosolic LC3 is converted to membrane-bound LC3 (LC3-II) in both hepatocytes and cardiac myocytes 24 hours after the start of starvation. Electron microscopic morphometry reveals that the volume densities of autophagosomes/autolysosomes and LDs increase in hepatocytes 24 and 48 hours after the onset of starvation, whereas autophagosomes and autolysosomes are rarely found in cardiac myocytes and the volume density of LDs is only counted and significantly increased in them [9]. The amounts of TAG in hepatocytes and cardiac myocytes significantly increase after the onset of starvation, whereas the increase in TAG amount is much lower in cardiac myocytes than in hepatocytes and continues until 24 hours. Moreover, LC3 is localized on the surface of LDs and LC3-II (lipidated form) is fractionated into a perilipin and ADRP (LD marker)-positive lipid fraction from the starved liver and cardiac myocytes, respectively. In fact, the surface of such LDs obtained from the LD fraction is labeled by gold particles showing the antigenicity of LC3. Taken together, these results indicate that the LC3 conjugation system is critically involved in lipid metabolism via LD formation.

4. Generation of LDs is inhibited in various LC3 mRNA-knockdown cultured cells

The cytoplasmic LD is an organelle that has a neutral lipid core with a single phospholipid layer. LDs are believed to be generated between the two leaflets of the endoplasmic reticulum (ER) membrane and to play various roles, such as high efficiency energy storage [16]. However, it remains largely unknown how LDs are generated or grow in the cytoplasm. As has been shown previously, the Atg conjugation system that is essential for autophagosome formation is involved in LD formation in hepatocytes and cardiac myocytes. This tendency has also been confirmed in white adipose tissue of conditional Atg7-knockout mice that show less production fat bodies in the tissue [17].

It has been shown that LDs temporally accumulate in the cultured cell lines during proliferation [22]. We have confirmed that LDs are produced in cultured cells seeded at a density of 70% confluency [11]. Accordingly, it has been shown that LDs that are stained with BODIPY are significantly augmented in PC12 cells 12 hours after the start of cultures, while immunosignals for LC3 are colocalized with BODIPY-positive LDs [11]. By immunoelectron microscopy, gold particles indicating LC3 are found on the surface of LDs in PC12 cells. Moreover, by cell fractionation the membrane type of LC3 is demonstrated in the perilipin-positive LD fraction. It still remains unknown whether LC3 itself is involved in LD formation. Since LC3 is a substrate of Atg7, cultured cell lines such as HeLa cells, PC12 cells, HepG2 cells, and Cos-1 cells, were examined to check the relationship of LC3 and LD formation. Expression of LC3 was suppressed by the method of RNA interference (RNAi), and it was found that LD formation is largely inhibited in these cells. TAG, a major component of LDs, is synthesized and degraded in LC3 mRNA-knockdown cells as well as in control cells. Interestingly, the potential for bulk protein degradation in the knockdown-cells is also evident in the control cells.

3T3 L1 cells, a progenitor cell line of adipocytes, accumulate LDs 12 hours after the start of cultures and LD formation is suppressed in the cells when mRNA of LC3 is knocked down [11]. Differentiation of L1 cells into adipocytes is confirmed by the mRNA expression of sterol regulatory element binding factor 1 (SREBF1) and peroxisome proliferator activated receptor γ (PPARγ), adipose specific proteins. It takes 6days until the L1 cells differentiate, and as the cells differentiate, it is found that the amount of LC3 mRNA also increases. In this differentiated situations, the surface of LDs in L1 cells is covered with perilipin and LC3. In LC3 mRNA-knockdown L1 cells, however, BODIPY-positive LDs largely disappear.

These findings indicate that LC3 is involved in the LD formation regardless of the bulk degradation, and that LC3 has two pivotal roles in cellular homeostasis mediated by autophagy and lipid metabolism.

5. Connection between autophagy and lipid metabolism

Recent studies provide supporting evidence for a connection between autophagy and lipid metabolism, both lipid storage and lipolysis. The involvement of autophagy in lipolysis of LDs in hepatocytes has been reported by the groups of Czaja and Cuervo, who showed that loss of Atg7 (Atg7[Flox/Flox]:albumin-Cre mice) results in accumulation of LDs in hepatocytes [18, 23-25]. Lipophagy, which is a form of autophagy that enwraps LDs by the isolation membrane has recently been considered important for the production of FFAs by degrading TAGs under acidic milieu and the FFAs produced fuel cellular rates of mitochondrial β-oxidation [18, 23-25]. This process, called lipophagy has recently been thought to function to regulate intracellular lipid stores, cellular levels of free lipids such as fatty acids and energy homeostasis [25]. On the contrary, as described in this chapter, it has been shown that loss of Atg7 (Atg7[Flox/Flox]:albumin-Cre mice) largely suppresses LD formation in hepatocytes and cardiac myocytes 24 hours after the start of starvation, although numerous LDs accumulate in normal hepatocytes and cardiac myocytes under the same conditions [9] (Fig. 1).

Electron microscopic analysis of wild-type mice by the former group shows LD-containing autophagosomes (lipophagosomes) under starvation conditions, although the latter group indicates that it is hard to see the presence of LDs enwrapped by double membranes even under starvation conditions, and that different from lipophagy, mitophagy can easily be found in hepatocytes if the mice are starved for 48 hours.

Two groups used different experimental approaches in their studies, which may underlie different conclusions. One critical point is to consider that Atg7[Flox/Flox]: albumin-Cre mice cause hepatomegaly and hepatitis with accumulation of abnormal organelles in hepatic cells [26].

Recently, metabolic contributions of amino acids released from liver by starvation-induced autophagy in adult animals using liver-specific Atg7-deficient mice have also been studied systematically [8]. That is, liver specific conditional Atg7-knockout mice (Atg7[Flox/Flox]:Mx1 mice) are generated by the different method from Atg7[Flox/Flox]:albumin-Cre mice [25]. To delete Atg7 from the liver, Cre expression in the liver was induced by intraperitoneal injection of polyinosinic acid-polycytidilic acid (pIpC), while complete deletion of the Atg7 protein in the liver was verified using immunoblotting analyses. Atg7-knockout mice are used 10 days after the injection of pIpC. No sign of hepatomegaly and hepatitis is observed within 2 weeks after the injection of pIpC. For synchronous induction of autophagy in the liver, mice previously fasted for 24 hours and preserve numerous LDs in hepatocytes, are fed by a pelleted laboratory diet for 2 hour (20:00–22:00) in the dark to suppress autophagic activity to a minimum [8]. The diet is then withdrawn and the mice are again starved. In separate experiments, it has been confirmed that the stomach and intestine are filled with a digested diet at the end of the 2 hour-feeding period. In this situation, hepatocytes become to change from the stage of lipogenesis and glycolysis to that of lipolysis and glycogenesis in hepatocytes. Electron microscopic examinations clearly show that in the control mice numerous LDs are continuously present in hepatocytes after 2 hour-feeding, and decrease during starvation [8]. Most LDs disappear in the liver after 24 hours of starvation, while glycogen granules increase in hepatocytes. In contrast, the number of autophagic vacuoles in the liver after 24 hour of starvation is increased dramatically. Such vacuoles and LDs are not observed in liver-specific Atg7-deficient mice [8].

These findings seem to indicate that LDs that accumulate in the liver during 24 hours fasting decrease in parallel with the following activation of autophagy after a 2 hour-feeding , but that loss of autophagy does not inhibit the disappearance of LDs in the liver. Thus another mechanism may operate in the removal of LDs from the liver that accumulated by starvation.

In adipose tissue, knock out of either Atg7 [9] or Atg5 [17] leads to reduced accumulation of lipid and impaired differentiation of adipocytes. The mutant white adipocytes are smaller with multiple LDs. It has been suggested that Atg5 deletion causes adipogenesis arrest at the later stages of mouse embryos. The possibility that a defect of Atg5-dependent Atg8 (LC3) lipidation and translocation to LDs of adipocytes may result in inefficient droplet fusion, which contributes to the defect of adipogenesis cannot be ruled out. The same authors observed reduced differentiation in Atg 5-deficient mouse embryonic fibroblasts (MEFs) model. An essential role of autophagy in lipid storage is also suggested in fat body cells of Drosophila. Knockdown of Atg1 or Atg6 led to small lipid droplets [27]. The paper has also indicated that Rab32 may regulate lipid storage by affecting autophagy [27]. Rab32 and several other Rabs have been found to affect the size of lipid droplets [27] and the mechanism of droplet fusion with atg8 (LC3) and Rab families awaits further studies. Thus, in adipocytes autophagy may contribute to LD formation and not significantly to lipolysis.

For hydrolysis of TAG in LDs of adipocytes, the molecular processes of lipolysis are becoming clear. Adipose tissue TAG lipase first acts on TAG to hydrolyze a fatty acyl chain [28]. Hormone sensitive lipase mediates the second step of lipolysis, diacylglycerol cleavage to monoacylglycerol. Finally monoacylglycerol lipase hydrolyzes the last side chain.

6. Concluding remarks

Lipid droplets (LDs) are key cellular organelles involved in lipid storage and mobilization. In non-adipocytes, LDs are small, mobile and interact with other cellular compartments. In contrast, adipocytes primarily contain very large, immotile LDs. The marked morphological differences between LDs in adipocytes and non-adipocytes suggest that key differences must exist in the manner in which LDs in different cell types interact with other organelles and undergo fusion and fission with other droplets. It has suggested that droplet fusion is dependent on microtubules, the motor protein dynein [29], Rab proteins [27] and the SNARE fusion machinery [31].

PAT proteins also target to the surface of LDs likely through different mechanisms. In adipocytes, TIP47 is found on smaller droplets and perilipin is found on larger LDs [31]. They may bind to different droplets associated proteins. Cleaved (and lipidated) LC3 demonstrated on isolated LDs from non-adipocytes [9] may be involved in LDs biology as well as PAT proteins (Figs. 2 and 3). Nakatogawa et al. [32] have shown that lipidated ATG8 (yeast homologue of LC3) mediates tethering between adjacent membranes of liposomes and stimulates membrane hemifusion, an event that may mimic expansion of the autophagosomal membrane during autophagy (Fig. 4). In LDs, homotropic interactions in adipocytes and non-adipocytes may involve a hemifusion type mechanism to facilitate lipid transfer (Fig.4). In adipocytes lipidated LC3 may also involved in growth of LDs (Fig. 4), as shown by morphological evidence that the mutant white adipocytes are smaller with multiple LDs [17].

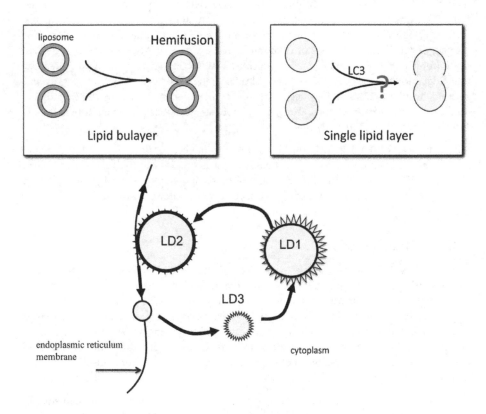

Figure 4. As shown by Murphy [16], a large LD3 that is attached to the endoplasmic reticulum may be degraded by ER-associated lipases and TAG may be re-synthesized. Newly formed LD1 would re-grow as LD2 in the cytoplasm. As has been shown by Nakatogawa et al. [32], bilayered liposomes with Atg8 undergo hemifusion (left upper), since Atg8 molecules are oligomerized with each other. If LC3 molecules are oligomelized on the surface of LDs (right upper), LDs would undergo fusion and grow in size.

It must be readdressed whether autophagy may affect lipid metabolism in a tissue- or cell-specific manner. If autophagy functions and regulates lipogenesis or lipolysis in a different way in non-adipocytes, it must be answered what molecular mechanisms work by which the autophagic machinery recognize LDs to enwrap for degradation or fuse with each other, and how they are regulated.

Author details

Yasuo Uchiyama and Eiki Kominami
Departments of Cell Biology and Neurosciences, and Biochemistry,
Juntendo University Graduate School of Medicine, Bunkyo-ku, Tokyo, Japan

Acknowledgement

This paper was supported by Grants-in-Aid for Scientific Research (B), (23390041) and partly for Scientific Research on Innovative Areas (23111004), and by MEXT-supported Program for the Strategic Research Foundation at Private Universities.

7. References

[1] Komatsu M, Waguri S, Chiba T, Murata S, Iwata J, Ueno Koike M, Uchiyama Y, Kominami E, Tanaka K. Loss of autophagy in the central nervous system causes neurodegeneration. Nature 2006 441: 880-884

[2] Komatsu M, Koike M, Ichimura Y, Uchiyama Y. Genetic mouse models for elucidation of autophagy-lysosomal systems in neurons under physiologic and pathologic conditions. In Ed. Zhenyu Yue, Charleen Chu: Autophagy of the nervous system – Cellular self-digestion in neurons and neurological diseases. World Scientific. (in press)

[3] Komatsu, M, Ueno, T, Waguri, S, Uchiyama, Y, Kominami, E, Tanaka, K. Constitutive autophagy: vital role in clearance of unfavorable proteins in neurons. Cell Death and Differentiation 2007 14:887-894

[4] Uchiyama Y, Shibata M, Koike M, Yoshimura K, Sasaki M. Autophagy - Physiology and Pathophysiology. Histochemistry and Cell Biology 2008 129: 407-420

[5] Mizushima M, Komatsu M. Autophagy: renovation of cells and tissues. Cell 2011 147:728-741

[6] Komatsu M, Waguri S, Koike M, Sou1 Y, Ueno T, Hara H, Mizushima N, Iwata J, Ezaki J, Murata S, Hamazaki J, Nishito Y, Iemura S, Natsume T, Yanagawa T, Uwayama J, Warabi E, Yoshida H, Ishii T, Kobayashi A, Yamamoto M, Yue Z, Uchiyama Y, Kominami E, Tanaka T. Homeostatic levels of p62 control cytoplasmic inclusion body formation in autophagy-deficient mice. Cell 2007 131: 1149-1163

[7] Mortimore GE, Pösö AR. Intracellular protein catabolism and its control during nutrient deprivation and supply. Annual Review of Nutrition 1987 7:539–564

[8]   Ezaki J, Matsumoto N, Takeda-Ezaki M, Komatsu M, Takahashi K, Hiraoka Y, Taka H,
      Fujimura T, Takehana K, Yoshida M, Iwata J, Tanida I, Furuya N, Zheng DM, Tada N,
      Tanaka K, Kominami E, Ueno T. Liver autophagy contributes to the maintenance of
      blood glucose and amino acid levels. Autophagy 2011 7:727-736

[9]   Shibata M, Yoshimura K, Furuya M, Koike M, Ueno T, Komatsu M, Arai H, Tanaka K,
      Kominami E, Uchiyama Y. The MAP1-LC3 conjugation system is involved in lipid
      droplet formation. Biochemical and Biophysical Research Communication 2009 382:
      419-423

[10] Gotoh K, Lu Z, Morita M, Shibata M, Koike M, Waguri S, DonoK, DokiY, Kominami E,
      Sugioka A, Monden M, Uchiyama Y. Participation of autophagy in the initiation of graft
      dysfunction after rat liver transplantation. Autophagy 2009 5:351-360

[11] Shibata M, Yoshimura K, Tamura H, Ueno T, Nishimura T, Inoue T, Sasaki M, Koike M,
      Arai H, Kominami E, Uchiyama Y. LC3, a microtubule-associated protein1A/B light
      chain3, is involved in cytoplasmic lipid droplet formation. Biochemical and Biophysical
      Research Communication 2010 393:274-279

[12] Koike M, Shibata S, Tadakoshi T, Gotoh K, Komatsu M, Waguri S, Kawahara N, Kuida K,
      Nagata S, Kominami E, Tanaka K, Uchiyama Y. Inhibition of autophagy prevents
      hippocampal pyramidal neuron death after hypoxic-ischemic injury. American Journal
      of Pathology 2008 172: 454-469

[13] Koike M, Shibata M, Waguri S, Yoshimura K, Tanida I, Kominami E, Gotow G,
      Peters C, Figura Kv, Mizushima N, Saftig P and Uchiyama Y. Participation of
      autophagy in storage of lysosomes in neurons from mouse models of neuronal
      ceroid-lipofuscinoses (Batten disease). American Journal of Pathology 2005
      167:1713-1728

[14] Martin S, Driessen K, Nixon SJ, Zerial M, Parton RG. Regulated localization of Rab18 to
      lipid droplets: effects of lipolytic stimulation and inhibition of lipid droplet catabolism.
      Journal of Biological Chemistry 2005 280:42325–42335

[15] Thiele C, Spandl J. Cell biology of lipid droplets. Current Opinion in Cell Biology 2008
      20:378–385

[16] Murphy DJ. The biogenesis and functions of lipid bodies in animals, plants and
      microorganisms. Progress in Lipid Research 2001 40:325–438

[17] Zhang Y, Goldmana S, Baergaa R, Zhaoa Y, Komatsu M, Jin S. Adipose-specific deletion
      of autophagy-related gene 7 (atg7) in mice reveals a role in adipogenesis. Proceedings
      of the National Academy of Sciences of the United States of America 2009 106:19860–
      19865

[18] Singh R, Cuervo AM. Lipophagy: Connecting autophagy and lipid metabolism.
      International Journal of Cell Biology 2012 2012:1-12

[19] Uchiyama Y, Asari A. A morphometric study of the variations in subcellular structures
      of rat hepatocytes during 24 hours. Cell and Tissue Research 1984 236:305-315

[20] Waguri S, Kohmura M, Kanamori S, Watanabe T, Ohsawa Y, Koike M, Tomiyama Y,
      Wakasugi M, Kominami E, Uchiyama Y. Different distribution patterns of the two

mannose 6-phosphate receptors in rat liver. Journal of Histochemistry and Cytochemistry 2001 49: 1397-1405

[21] Mizushima N, Yamamoto A, Matsui M, Yoshimori T, Ohsumi Y. In vivo analysis of autophagy in response to nutrient starvation using transgenic mice expressing a fluorescent autophagosome marker. Molecular Biology of Cell 2004 15:1101–1111.

[22] Swinnen JV, Brusselmans K, Verhoeven G. Increased lipogenesis in cancer cells: new players, novel targets, Current Opinion in Clinical Nutrition and Metabolic Care 2006 9:358–365.

[23] Singh R, Kaushik S, Wang Y, Xiang Y, Novak I, Komatsu M, Tanaka K, Cuervo AM, Czaja MJ. Autophagy regulates lipid metabolism. Nature 2009 458:1131-1135

[24] Singh R, Xiang Y, Wang Y, Baikati K, Cuervo AM, Luu YK, Tang Y, Pessin JE, Schwartz GJ, Czaja MJ. Autophagy regulates adipose mass and differentiation in mice. Journal of Clinical Investigation 2009 119:3329-39

[25] Liu K, Czaja MJ. Regulation of lipid stores and metabolism by lipophagy. Cell Death and Differentiation 2012 doi:10.1038/cdd.2012.63

[26] Komatsu M, Waguri S, Ueno T, Iwata J, Murata S, Tanida I, Ezaki J, Mizushima N, Ohsumi Y, Uchiyama Y, Kominami E, Tanaka K, Chiba T. Impairment of starvation-induced and constitutive autophagy in Atg7-deficient mice. Journal of Cell Biology 2065 169: 425–434

[27] Wang C, Liu Z, Huang X. Rab32 Is Important for Autophagy and Lipid Storage in Drosophila. PLos ONE 2012 7e32086

[28] Zimmermann R, Strauss JG, Haemmerle G, Schoiswohl G, Birner-Gruenberger R, Riederer M, Lass A, Neuberger G, Eisenhaber F, Hermetter A, and Zechner R. Fat mobilization in adipose tissue is promoted by adipose triglyceride lipase. Science 2004 306;1383–1386

[29] Mendis-Handagama SMLC, Aten RF, Watkins PA, Scallen TJ, Berhman RH. Peroxisomes and sterol carrier protein-2 in luteal cell steroidogenesis: a possible role in cholesterol transport from lipid droplets to mitochondria. Tissue and Cell 1095 27+483-490

[30] Boström P, Rutberg M, Ericsson J, Holmdahl P, Andersson L, Frohman M, Borén J, Olofsson S. Cytosolic lipid droplets increase in size by microtubule-dependent complex formation. Arteriosclerosis, Thrombosis and Vascular Biology 2005 25:1945–1951.

[31] Boström P, Andersson Rutberg LM, Perman Lidberg JU, Johansson BR, Fernandez-Rodriguez J, Ericson j, Nilsson T, Boren j, Olofsson SO. SNARE proteins mediate fusion between cytosolic lipid droplets and are implicated in insulin sensitivity. Nature Cell Biology 2007 9:1286–1293

[32] Wolins N, Quaynor B, Skinner J, Schoenfish M, Tzekov A, Bickel P. S3-12, adipophilin, and TIP47 package lipid in adipocytes. Journal of Biological Chemistry 2005 280:19146–19155

[33] Nakatogawa H, Ichimura Y, Ohsumi Y. Atg8, a ubiquitin-like protein required for autophagosome formation, mediates membrane tethering and hemifusion. Cell 2007 130:165–178

Metabolism of Short Chain Fatty Acids in the Colon and Faeces of Mice After a Supplementation of Diets with Agave Fructans

Alicia Huazano-García and Mercedes G. López

Additional information is available at the end of the chapter

1. Introduction

1.1. Fructans

Most plants store starch or sucrose as reserve carbohydrates, but approximately 12-15% of higher plants (representing more than 40,000 species) synthesizes fructans as their main source of carbohydrates [1]. Fructans are found naturally in plants as a heterogeneous mixture of different polymerization degrees, they are a polydisperse mixture. Among plants that store fructans, many are economically important, due to its content of fructans, as it is the case of chicory (*Cichorium intybus*), agave (*Agave spp.*), artichoke (*Cynara scolymus*), dahlia (*Dahlia variabilis*), garlic (*Allium sativum*) and wheat (*Triticum asetivum*) [2, 3]. Five different groups of fructans have been found in nature and distinguished according to the type of linkage between fructose units and the position of the glucose moiety within the structure. These groups consist of inulins, neoseries inulins, levans, neoseries levans and graminans. Inulins consist of a linear β(2-1) linked fructosyl chain; neoseries inulins are composed of two linear β(2-1) linked fructosyl chains, one bound to the fructosyl residue of the sucrose, the other bound to the glucosyl residue of the same sucrose molecule; levans consist a of linear β(2-6) linked fructosyl chain; neoseries levans are composed of two linear β(2-6) linked fructosyl chains, one bound to the fructosyl residue of the sucrose, the other bound to the glucosyl residue and graminans which present both linkages, β(2-1) and β(2-6) links to the fructose moiety of sucrose [4].

Currently, inulins are extracted from chicory roots, containing fructose chains having a degree of polymerization (DP) from 3 to 60 [2] (Figure 1a). The chemical or enzymatic (endoinulinases) hydrolysis of inulins produces inulins of shorter DP (DP<10), these are called fructooligosaccharides (FOS) [5, 6].

Mexico is considered the origin center of evolution and diversification of the *Agave genus*, since a large number of agave species are found in its territory. The *Agave genus* includes approximately 166 species and is the largest *genus* among the *Agavaceae* family that consists of 9 genera and approximately 293 species [7, 8]. The agave plants have the ability to grow in extremely dry-hot environments, where sometimes this plant is the predominant or exclusive flora in that type of a geo-climatic zone, however, they can also be found in diverse ecosystems, such as productive highlands and elevated humidity [9]. These plants present a crassulacean acid metabolism (CAM) and their principal photosynthetic products are fructans [10], fructans are synthesized and stored in the stems of agave plants. Agave is the most exploited genus and economically important as the raw materials are used on the production of alcoholic beverages such tequila (*A. tequilana*) and mezcal (*A. angustifolia, A. potatorum, A. cantala, A. duranguensis*, to mention some) in Mexico. *A. angustifolia* is an endemic plant that grows in different states of Mexico; however the main producing states are Oaxaca and Sonora. Fructans in *A. angustifolia* from Oaxaca represent more than 85% of total water soluble carbohydrates in the plant, with an estimated DP of 32 [11].

Agave fructans posses a molecular structure compose of a complex mixture containing highly branched molecules with β(2-1) and β(2-6) linkages, as well as internal and external glucose units, due to the existence of both types of glucose, agave fructans have been classified as graminans (external glucose) and agavins (internal glucose) [11] (Figure 1b).

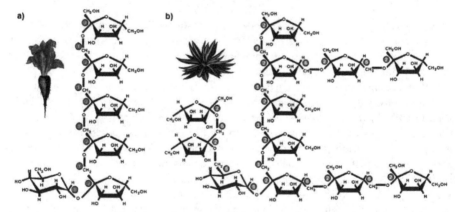

Figure 1. Schematic representation of the main structural differences between a) chicory fructans, "inulins" and b) agave fructans, "agavins".

All fructans are considered prebiotics molecules that serve as a substrate for the gut microbiota [6, 12-15]. A prebiotic is an ingredient selectively fermented by probiotics (*bifidobacteria* and *lactobacilli*) that induces specific changes on the composition and/or activity of the gastrointestinal microbiota, conferring benefits upon the host well-being and health in general [16, 17]. The fermentation of fructans in the colon generates short chain fatty acids (SCFAs). SCFAs formation is an important event since it favors the maintenance and the development of beneficial microbiota as well as the colonic epithelial cells [16].

Metabolism of Short Chain Fatty Acids in the Colon and Faeces of Mice After a
Supplementation of Diets with Agave Fructans

135

1.2. Short chain fatty acids (SCFAs)

The gastrointestinal tract is an extremely complex ecosystem containing about 10^{11} CFU (colony forming units) of bacteria per gram of intestinal content. This large population of bacteria plays a key role in the nutrition and health of the host [18]. The colonic microbiota ferments organic material that cannot be digested otherwise by the host in the upper gut. These include resistant starch, non-digestible carbohydrates (fructans) as well as some proteins and amino acids [19]. The main products of fructans metabolism in the colon are linear SCFAs, mostly acetate ($C_{2:0}$), propionate ($C_{3:0}$) and butyrate ($C_{4:0}$) [19-21] (Figure 2). However, other fermentation products may be lactate, succinate as well as ethanol [6], which are sometimes only intermediates in the global process of carbohydrates fermentation by the microbiota, and are metabolized in varying degrees to SCFAs by interactions and/or collaboration of present bacteria in the ecosystem, so that generally do not accumulate to any significant extent in the colon [22]. Fructans fermentation also produces a few gases as CO_2, CH_4, H_2 and additionally heat [19, 23]. The presence of both, non-digestible carbohydrates and SCFAs in the colon can positively alter the colonic physiology drastically [24]. Various studies on microbial population have shown that SCFAs production is in the order of $C_{2:0} > C_{3:0} > C_{4:0}$ in a molar ratio of approximately 60:20:20 mainly in the proximal and distal colon [19, 25]. An increased in SCFAs synthesis also creates a more acidic environment in the gut, which is important *in vivo* in terms of colonization resistance against pathogens [18, 20]. The production of SCFAs is affected by many factors, including the source of substrate [26], in particular, the chemical composition of the fermentable substrate, the amount of substrate available, its physical form (e.g. particle size, solubility, association with undigestible complexes such as lignin) [27], the bacterial species composition of the microbiota [12], ecological factors (competitive and cooperative interactions between different groups of bacteria) and intestinal transit time [25]. The gut of mice comprises four sections: caecum, proximal, transverse (medial) and distal colon. The caecum and proximal colon are the main sites where fermentation is carried out, given the number of bacteria and the availability of substrate, because as it moves through the intestine toward the distal colon, there is a lower concentration of water as well as a depletion of carbohydrates and increased pH [22]. SCFAs are rapidly absorbed in the caecum and colon being excreted in the faeces only from 5% to 10% of them [24]. The major SCFAs ($C_{2:0}$, $C_{3:0}$ and $C_{4:0}$), are absorbed at comparable rates in different regions of the colon. Once absorbed, SCFAs are metabolized at three major sites in the body: 1) cells of the caecum-colonic epithelium that use $C_{4:0}$ as a major substrate for maintenance-energy; 2) liver cells that metabolize residual $C_{4:0}$ and $C_{3:0}$ used for gluconeogenesis and 50% to 70% of $C_{2:0}$ is also taken up by the liver; and 3) muscle cells that generate energy from the oxidation of residual $C_{2:0}$ [3].

1.2.1. Acetic ($C_{2:0}$), propionic ($C_{3:0}$) and butyric ($C_{4:0}$) acids

$C_{2:0}$ is the principal SCFA produced in the colon, this is readily absorbed and transported to the liver, and therefore is less metabolized in the colon [26]. The presence of acetyl-CoA synthetase in the cytosol of adipose and mammary glands allows the use of $C_{2:0}$ for lipogenesis once it enters the systemic circulation [24]. $C_{2:0}$ is the primary substrate for cholesterol synthesis. In the host, it may be absorbed and utilized by peripheral tissues also [29].

On the other hand, $C_{3:0}$ is produced via two main pathways: 1) by fixation of CO_2 to form succinate, which is subsequently decarboxylated (the "dicarboxylic acid pathway") and 2) forms lactate and acrylate (the "acrilate pathway") [30]. $C_{3:0}$ is also a substrate for hepatic gluconeogenesis and it has been reported that this acid inhibits cholesterol synthesis in hepatic tissue [31, 32]. The ratio of $C_{3:0}$ to $C_{2:0}$ in the colon is relevant since it lowers cholesterol synthesis coming from the $C_{2:0}$ pathway [32].

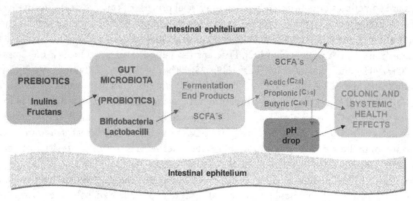

Figure 2. General events taken place in the large intestine. Prebiotics are the specific food for probiotics which ferment them to produce short chain fatty acids (SCFAs) to improve the host health.

Finally, $C_{4:0}$ is the preferred fuel by the colonic epithelial cells but also plays a major role in the regulation of cell proliferation and differentiation [19]. It is the most important SCFA in colonocytes metabolism, where 70% to 90% of $C_{4:0}$ is metabolized by the colonocytes. $C_{4:0}$ is used preferentially over $C_{3:0}$ and $C_{2:0}$ in a ratio of 90:30:50 [26]. Approximately 95% of the $C_{4:0}$ produced by colonic bacteria is transported across the epithelium, but concentrations in portal blood are usually undetectable as a result of a rapid utilization [33]. $C_{4:0}$ production might also occur through the use of other fermentation products such as $C_{2:0}$ or lactate that can act as precursors of $C_{4:0}$.

1.2.2. Production of SCFAs in vitro

In vitro SCFAs production can be measured using pure cultures of selected bacteria species of faecal slurry and some prebiotics (undigestible carbohydrates, for instance).

a.   Inulins

Studies *in vitro* using faecal inocula incubation have shown that different substrates (prebiotics) yield various SCFAs patterns; Van de Wiele et al. [34] compared the fermentation of FOS and inulin *in vitro* with faecal inoculum, observing that FOS and inulin

Metabolism of Short Chain Fatty Acids in the Colon and Faeces of Mice After a
Supplementation of Diets with Agave Fructans

137

increased the production of SCFAs by about 30%. This increment in SCFAs production was attributed to an increase mainly in $C_{3:0}$ and $C_{4:0}$ acids. Where inulin showed a higher production of $C_{3:0}$ acid (almost 2-fold) than FOS. The authors concluded that these differences correlated well with the structural differences, FOS has a short DP and inulin a long DP. Stewart et al. [35] analyzed the fermentation of inulins with three different chain lengths (short, medium and long). These researchers found that short inulins were rapidly fermented and produced higher concentrations of $C_{4:0}$ compared with other inulins, hence, chain length is an important factor on the fermentation patterns of SCFAs (Table 1). Rycroft et al. [36] performed a comparative evaluation of different prebiotics *in vitro*, they observed that all the used substrates presented an increased concentration of $C_{2:0}$ acid. Inulins generated the highest concentrations of $C_{3:0}$ acid, but a mixture of FOS and inulin showed the highest production of $C_{4:0}$ acid than the other prebiotics by themselves. The differences in SCFAs patterns in these studies may be also attributed to differences in bacteria species present in the faecal inocula or fermentation process used.

Sample	DP (mean)	Time	$C_{2:0}$*		$C_{3:0}$*		$C_{4:0}$*	
			Mean	SEM	Mean	SEM	Mean	SEM
A	< 5	12 h	22.3ᵃ	0.50	5.3ᵃ	0.20	12.8ᵃ	0.20
		24 h	16.9ᵃᵇ	0.10	2.1	0.10	9.1ᵃ	0.30
B	= 4.8	12 h	20.2ᵃᵇ	1.50	4.1ᵇᶜ	0.02	10.2ᵇ	0.50
		24 h	15.1ᵇᵈ	0.03	1.8	0.20	9.0ᵃ	0.40
C	< 10	12 h	26.8ᵃ	0.60	4.7ᵃᵇ	0.20	12.8ᵃ	0.30
		24 h	20.9ᶜ	0.80	2.7	0.30	9.6ᵃ	0.10
D	≈ 10	12 h	20.0ᵃᵇ	0.10	2.1ᵈ	0.10	9.3ᵇ	0.20
		24 h	14.4ᵈ	0.40	1.8	1.00	8.7ᵃ	0.30
E	> 10	12 h	19.9ᵃᵇ	1.30	3.6ᶜ	0.10	9.2ᵇ	0.02
		24 h	25.1ᵃ	0.04	3.8	0.10	9.3ᵃ	0.10
F	> 20	12 h	14.0ᵇ	2.10	0.8ᵉ	0.20	5.4ᶜ	0.10
		24 h	18.7ᵃ	0.20	2.4	0.10	6.8ᵇ	0.10

*Concentration μmol/mL. Values with different letter within a time point are statistically different from each other (P<0.05). [35, with modifications].

Table 1. Concentration of acetic ($C_{2:0}$), propionic ($C_{3:0}$) and butyric ($C_{4:0}$) acids obtained after 12 and 24 h *in vitro* fermentation batch with inulins of different chain length.

b. Agave fructans

Urías-Silvas & López [37] analyzed the prebiotic potential of fructans extracted from five different species of *Agave* spp. grown in different regions of Mexico, *Dasylirion* sp. and commercial inulins, using strains of *bifidobacteria* and *lactobacilli*. These researchers found that branched fructans from *Dasylirion* (DSC) with a DP range from 3 to 20 and *A. tequilana* from the state of Guanajuato (ATG) with a DP between 3 and 22, stimulated better the growth of both bacteria genera in MRS medium (Figure 3). Moreover, the major SCFAs fermentation product, were acetic, formic and lactic acids, wherein the proportions of the acids varied depending on the prebiotic type used by the different bacteria. Figure 4 shows the fermentation products only for the two agave fructans (DSC and ATG) and commercial inulins (RSE and RNE) that better stimulated the growth of bacteria. In general, in figure 4 it can be observed that the branched fructans (agavins) were able to produce more acids than the linear fructans (inulins).

Santiago-García & López [38] studied the *in vitro* prebiotic effect of fructans from *A. angustifolia* of long-DP, short-DP and three combinations of them. The growth rate of *bifidobateria* and *lactobacilli* strains with *A. angustifolia* fructans was compared with commercial inulins (Raftiline and Raftilose). The authors observed that agave fructans stimulated the growth of *bifidobacteria* and *lactobacilli* more efficiently (2-fold) that commercial inulins, either long- or short-DP. They also reported that short-DP fructans in the mixtures highly influenced the rate of fermentation by probiotc bacteria. In this work, the main fermentation product in all treatments was $C_{2:0}$ acid. Moreover, Gomez et al. [39] compared the growth of *bifidobacteria* and *lactobacilli* on a complex faecal microbiota *in vitro* using fructans extracted from *A. tequilana Weber* var. azul and different commercial inulins. Their results indicated no significant differences among the growth of both bacteria genera with the different fructans used. With regard to the total SCFAs production by agave fructans and inulins was very similar. $C_{2:0}$ acid was the most prevalent SCFA in all treatments, only agave fructans and inulin of long-DP produced significantly higher amounts of $C_{3:0}$, however, there were no significant differences between the different fructans used.

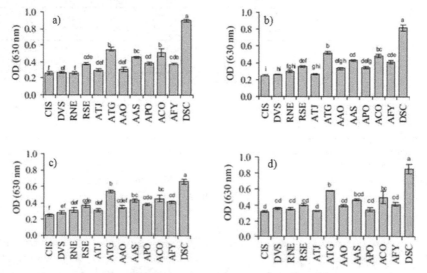

Figure 3. Effect of different fructans on the growth of **a)** *B. adolescentis;* **b)** *B. infantis;* **c)** *L. paracasei* and **d)** *L. rhamnosus* incubated anaerobically at 37°C in the presence of 10 g of fructan/L. OD, Optical density; CIS, *Cichorium intybus;* DVS, *Dahlia variabilis;* RNE, Raftiline; RSE, Raftilose; ATJ, *A. tequilana* Jalisco; ATG, *A. tequilana* Guanajuato; AAO, *A. angustifolia* Oaxaca; AAS, *A. angustifolia* Sonora; APO, *A. potatorum* Oaxaca; ACO, *A. cantala* Oaxaca; AFY, *A. fourcroydes* Yucatán; DSC, *Dasylirion* sp. Chihuahua.

1.2.3. Production of SCFAs in vivo

a.   Inulins

Nilsson and Nyman [40] evaluated the formation of SCFAs in the hindgut of rats fed with lactulose, lactitol, FOS and inulins of different DP and solubility. The major acids formed were $C_{2:0}$, $C_{3:0}$ and $C_{4:0}$. The highest levels of $C_{3:0}$ acid were found in caecum and proximal and distal colon of rats fed with inulins, whereas the highest levels of $C_{4:0}$ acid were found in caecum and proximal and distal colon of rats fed with FOS. The authors concluded that the DP and solubility of the used prebiotics were of great importance on SCFAs production.

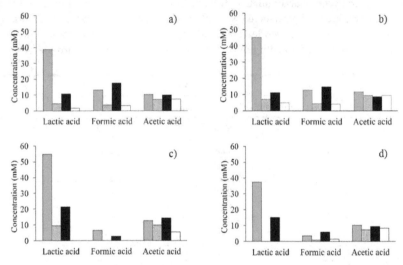

Figure 4. Concentration of short chain fatty acids generated by *bifidobacteria* and *lactobacilli* from the fermentation of *Dasylirion* sp. (DSC;), Raftilose (RSE;), *A. tequilana* GTO. (ATG;) and Raftiline (RNE;). **a)** *B. adolescentis*; **b)** *B. infantis*; **c)** *L. paracasei* and **d)** *L. rhamnosus*.

Similar results were obtained by Licht et al. [41] who fed rats with different dietary carbohydrates. These authors concluded that $C_{3:0}$ acid concentrations reached statistical significance in animals fed with inulins, whereas the concentration of $C_{4:0}$ acid was significantly higher in animals receiving FOS. Klessen et al. [42] also determined the production of SCFAs in the caecum and colon of germ-free rats associated with contents of human faecal, the rats were fed with inulins of different chain lengths (FOS, inulins and a mixture of FOS-inulins). They observed that FOS produced the greatest amount of $C_{2:0}$ acid in the colon of the rats whereas inulins increased the concentration of $C_{3:0}$ acid in the caecum of the animals that consumed this diet. Moreover, FOS, inulins and the mixture of FOS-inulins increased the amount of $C_{4:0}$ acid in the caecum and colon of the rats fed with the mixture regard to animals fed with standard diet. The authors concluded that the type of diet and the fermentation site in the colon affected the concentration of SCFAs (Table 2). In another work, Levrat et al. [43] fed rats with different percentages of inulins (5, 10 and 20%), finding that $C_{2:0}$ acid production was significantly lower in rats fed with 20% inulin diet. Moreover, all percentages of inulin increased the levels of $C_{3:0}$ acid in the caecum of the rats; the highest concentration was found in animals that consumed the 10% inulin diet whereas $C_{4:0}$ acid concentration was markedly enhanced in all supplemented diets in spite of the inulin percentage used. In another study, the same authors fed rats with 10% of inulin, they found a higher concentration of $C_{3:0}$ acid in the portal vein as well as a significant decrease in plasma cholesterol levels of the rats fed with this diet with regard to animals that consumed the standard diet [44]. On the other hand, a study carried out using obese rats that received a diet supplement with inulin, a two-fold greater $C_{3:0}$ concentration in the portal vein and a decrement on triglyceride accumulation in the liver of these animals was observed [45]. A similar result was seen in hamsters fed with different percentages of inulins (8, 12 and 16%). Plasma cholesterol and triglyceride concentrations were significantly lower with all the percentages of inulins studied with respect to hamsters fed with the standard diet [46].

Acid	Segment	Diet				Pooled SEM
		Standard	FOS	Inulin	Mix (FOS-Inulin)[2]	
$C_{2:0}$[1]	Caecum	49.8	55.4	45.9	51.2	1.3
	Colon	41.1	50.8*	42.5	46.7	1.2
	Faeces	37.4	35.1	35.6	27.6*	1.3
$C_{3:0}$[1]	Caecum	21.1	22.8	32.5*	19.1	0.8
	Colon	18.2	17.8	21.3	16.8	0.7
	Faeces	14.7	15.5	14.4	15.8	0.4
$C_{4:0}$[1]	Caecum	13.4	21.3*	25.4*	28.0*	1.3
	Colon	9.3	18.6*	18.1*	22.3*	1.2
	Faeces	7.1	11.4	13.6*	15.7*	0.9

Mean values n=6; [1]Concentrations [µmol/g wet wt]; [2]Mix FOS-Inulin (1:1 w/w). Mean values were significantly different from those of the standard diet group: *P<0.05. [42, with modifications].

Table 2. Production of short chain fatty acids in the caecum, colon and faeces in rats associated with human faecal contents and fed with inulins of different chain lengths.

b.   Agave fructans

To date, there are no published reports on the production of SCFAs *in vivo* using agave fructans of any species. However, the physiological effects of fructans extracted from *A. tequilana* Gto. (ATG) and *Dasylirion* sp. (DSC) have shown that agave fructans positively impact some lipid metabolic molecules such triglycerides and cholesterol on serum of mice fed with diets supplemented with these types of fructans (Figure 5) they also affect glucose levels [47]. These effects were attributed to the production of $C_{3:0}$, which is largely produced through the fermentation of all fructans.

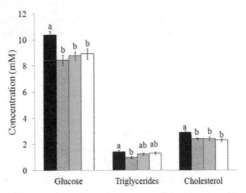

Figure 5. Concentration of glucose, triglycerides and cholesterol in mice after consumption of a standard diet (STD;■) or diet supplemented with fructans: chicory (RSE;▨); *A. tequilana* Gto. (ATG;▨) and *Dasylirion* sp. (DSC;□). Mean values n=8 with their standard errors of the mean for each parameter measured. Mean values with different letters were significantly different (P≤0.05).

In another study with fructans of *Dasylirion* sp. (DSC) and commercial inulin Raftilose (RSE), García-Pérez [48] reported that the diets supplemented with fructans had a beneficial effect on the concentration of glucose and cholesterol in blood of the portal vein of mice. Glucose concentrations were significantly lowered by 22 and 27% in mice fed DAS and RSE diets with respect to mice fed a standard diet. Cholesterol concentrations were also reduced by 20% in animals receiving DSC and 14% in mice fed RSE diet. However, levels of triglycerides were not significantly modified by any treatment. In this same study, SCFAs were determined only in faeces (Figure 6). In general, mice fed diets supplemented with DSC present higher amount of $C_{2:0}$, $C_{3:0}$ and $C_{4:0}$ acids in their faeces. Faecal concentrations of SCFAs are not of course the best way to measure the production rates since large proportion of SCFAs is taken up by the colonic mucosa. Nevertheless, faecal levels of SCFAs are a good marker or indicator of the differences on SCFAs taken place in the gut of mice that consumed fructans diets.

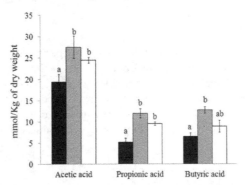

Figure 6. Concentration of SCFAs in faeces of mice fed with a standard diet (STD;■) or diet supplemented with Raftilose (RSE;▨) or *Dasylirion* spp. (DSC;□). Mean values n=8 with their standard errors of the mean. Mean values with different letters were significantly different (P≤0.05).

With all the above, we decided to run an *in vivo* experiment feeding mice with *Agave angustifolia* fructans and evaluating the formation of SCFAs in caecum, colonic sections and faeces, as well as the pH drop in all these areas of the gut.

2. Materials and methods

2.1. Animals and diets

Twenty-four male C57BL/6J mice of 12 weeks old at the beginning of the experiment were obtained from the animal facilities of CINVESTAV-Zacatenco (Mexico). The mice were housed in a temperature and humidity controlled room with a 12 h light-dark cycle. They were divided into three groups (eight mice per group) according to diet. Mice were acclimated for 7 days, having free access to a pelleted 5053 standard diet (Laboratory Rodent Diet, USA) and water. During the experimental period (6 weeks), STD mice group were fed with 5053 standard diet, whereas inulins-RNE and agavins-AAO mice groups received a diet prepared by mixing 90 g of 5053 standard diet with 10 g of Raftiline or fructans from *A. angustifolia* (AAO). All diets were made by Laboratory Rodent Diet and were available *ad libitum* to mice.

2.2. Chemicals

Agavins were extracted in our laboratory as described by López et al. [10]. Briefly, one hundred grams of milled *Agave angustifolia* stems were extracted twice with 100 ml of 80% v/v ethanol with continuous shaking for 1 h at 55 °C. The sample was filtered and the plant material re-extracted with 100 ml of water for 30 min at 55 °C. The supernatants were mixed; chloroform was used to eliminate the organic fraction. The aqueous phase was concentrated by rotary evaporation under reduced pressure. The sample was dried using a spray dryer and stored in a humidity-free container. RNE were from Beneo Orafti. The average degree polymerization for agavins is 32[11] and for RNE >23.

2.3. Faeces and blood samples

Faeces collection was performed once a week during the experimental period to evaluate the SCFAs. On day 45, mice were anaesthetized by intra-peritoneal injection of sodium pentobarbital solution. Portal vein blood samples were collected in EDTA tubes; after centrifugation for 10 minutes at 2500 r.p.m., plasma was stored at -80 °C. The concentration of serum triglycerides, cholesterol and glucose was measured using kits coupling enzymatic reaction and spectrophotometric detection of reaction end-products (BioVision).

2.4. pH and SCFAs

Segments of the caecum and proximal, medial and distal colon were immediately excised. Caecal and colonic contents of each section were collected in tubes and frozen at -80 °C. The pH measurements were made using a microelectrode (PHR-146, Lazar Research

Laboratories, Inc.). Analysis of SCFAs was carried out by gas chromatography and flame ionization detection as described by Pietro Femia et al. [49] with some modifications. Briefly, 0.05 g of caecal and faecal contents were acidified with 0.05 ml of sulfuric acid and SCFAs were extracted by shaking with 0.6 ml of diethyl ether and subsequent centrifugation at 14000 r.p.m. for 30 s. One microliter of the organic phase was injected immediately into the capillary column (Nukol) of the gas chromatograph coupled to a flame ionization detector. The initial temperature was 80 °C and the final temperature was 200 °C. Nitrogen was used as carrier gas and the quantification of the samples was carried out using calibration curves for $C_{2:0}$, $C_{3:0}$ and $C_{4:0}$ acids. A standard curve for each acid was done for their quantitation in the samples.

2.5. Statistical analysis

Results are expressed as mean values with their standard errors of the mean. Statistical differences between groups were evaluated using one-way ANOVA followed by a Tukey test using GraphPad Prism version 5 for Windows. P<0.05 was regarded as statistically significant.

3. Results

3.1. Feed intake and body weight

The intake of all mice independently of the diet fed ranged between 3.3 and 4.2 g/d with an average of 3.7 g/d, it is worth to mention that the intake fluctuated weekly throughout the study. The feed intake was 9% lower for the AAO group compared to the STD and RNE diets. Mice fed with the diet supplemented with RNE ate 10% more food than even the STD group. Initial body weights ranged from 21.4 to 24.4 g with final body weights ranging between 24.3 and 25.9 g. No significant differences among all groups were noted in body weight even though mice fed AAO reduced their intake.

3.2. Production of SCFAs and pH drop

The total production of SCFAs was greater for the group of mice fed with AAO in the caecum and proximal and medial colon. However in the distal colon, the production of SCFAs were not significantly different among supplemented diets but it did with the STD diet (Table 3).

$C_{2:0}$ was the most abundant acid formed in the caecum and colon of all mice followed by $C_{3:0}$ and $C_{4:0}$ acid. The concentrations of $C_{2:0}$ acid were significantly higher in the caecum and the first two sections of the colon (proximal and medial) in mice fed with AAO diet compared to RNE or STD groups. However, in the distal colon there were no significant differences on the production of $C_{2:0}$ acid between groups of mice fed fructans (Figure 7a). The higher concentration of $C_{3:0}$ acid was found in the caecum of mice fed with AAO diet. This increment was significant with regard to RNE but not for the STD diet. In the proximal and medial colon $C_{3:0}$ acid production was greater for mice fed AAO, but these enhancements were not significant. Interestingly, in the distal colon of mice fed fructans (AAO and RNE) the enhancement was significantly for the production of $C_{3:0}$ acid (Figure 7b). The concentration of $C_{4:0}$ acid increased approximately 24% in the caecum of mice fed with RNE diet. This enhanced was significant with regard to AAO but not for the STD diet. Finally, an increment of $C_{4:0}$ acid was observed in the medial and distal colon, this change was significant in mice fed with AAO and RNE diets compared to mice fed with the STD diet (Figure 7c).

SCFAs changes were confirmed when the pH was measured in all the same samples. The mice fed AAO diet showed a pronounced pH drop in the caecum and all sections of the colon. The group of mice fed RNE showed significant pH drop only in the medial and distal colon and the pH of the mice fed a STD diet did not change significantly in any sections of the gut (Figure 8). The pH drop changes positively correlated with the total production of SCFAs (Table 3).

Section	STD*		RNE*		AAO*	
	Mean	SEM	Mean	SEM	Mean	SEM
Caecum	73.93[a]	1.98	70.90[a]	2.08	83.48[b]	1.54
Proximal colon	71.98[a]	2.11	76.93[a]	2.62	106.92[b]	4.61
Medial colon	51.44[a]	1.25	64.26[b]	2.59	107.25[b]	2.41
Distal colon	28.58[a]	0.86	49.65[b]	2.91	66.84[b]	2.97

*[mmol/Kg of fresh weight]. Mean values n=8 with their standard errors of the mean. [a,b] Mean values with unlike superscript letters were significantly different (P≤0.05).

Table 3. Concentration of total short chain fatty acids production in the caecum and the three sections of the colon in mice fed a standard (STD) diet or diet supplemented with Raftiline (RNE-inulin) or *Agave angustifolia* Oax. (AAO-agavins).

Metabolism of Short Chain Fatty Acids in the Colon and Faeces of Mice After a
Supplementation of Diets with Agave Fructans

145

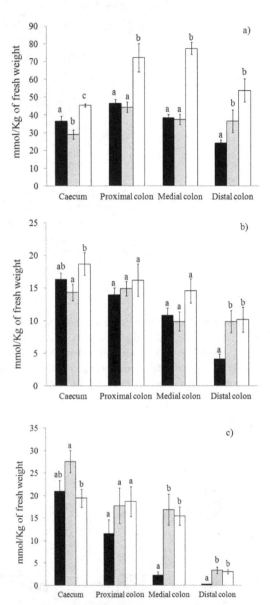

Figure 7. Concentrations of SCFAs in the caecum and the three sections of the colon of mice fed with a
standard diet (STD; ■) and supplemented diets containing Raftiline (RNE; ▨) and *A. angustifolia* Oax.
(AAO; ☐). A) Acetic acid ($C_{2:0}$); B) Propionic acid ($C_{3:0}$); and C) Butyric acid ($C_{4:0}$). Mean values n=8 with
their standard errors of the mean for each parameter measured. Mean values with different letters were
significantly different (P≤0.05).

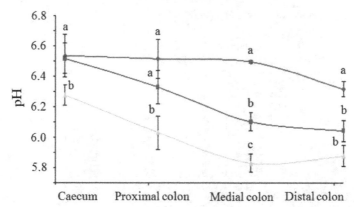

Figure 8. pH values in the caecum and the three sections of the colon of mice fed a standard diet (STD; ■) or diet supplemented with Raftiline (RNE; ■) or *A. angustifolia* Oax. (AAO;). Mean values n=8 with their standard errors of the mean. Mean values with different letters were significantly different (P≤0.05).

3.3. SCFAs in the faeces

The analyses of SCFAs in the collected faeces showed that $C_{2:0}$ acid was again the most abundant acid in the faeces of all mice followed by $C_{3:0}$ and $C_{4:0}$ acids. However, the amounts of $C_{2:0}$ and $C_{3:0}$ acids excreted in the faeces were not affected significantly by any dietary treatment. Surprisingly, only the mice fed with agavins (AAO) or inulin (RNE) diets, showed a significant increment on the amount of $C_{4:0}$ acid compared with the STD diet (Figure 9).

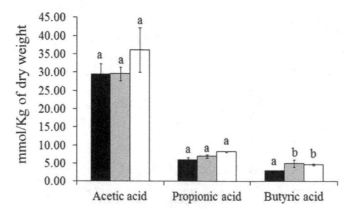

Figure 9. Concentration of acetic ($C_{2:0}$), propionic ($C_{3:0}$) and butyric ($C_{4:0}$) acids excreted in the faeces of mice fed a standard diet (STD;■) or diet supplemented with Raftiline (RNE;□) or *A. angustifolia* Oax. (AAO;□). Mean values n=8 with their standard errors of the mean. Mean values with different letters were significantly different (P≤0.05).

Metabolism of Short Chain Fatty Acids in the Colon and Faeces of Mice After a
Supplementation of Diets with Agave Fructans

147

3.4. Plasma glucose, triglyceride and cholesterol

Besides SCFAs some other physiological parameters were determined in the plasma of all mice groups, among them glucose, triglycerides and cholesterol. Glucose concentrations of mice fed AAO and RNE diets were significantly lowered by 12% and 17% respectively when compared to the STD diet. On the other hand, triglycerides concentrations of the supplemented groups compared with the STD were reduced by 37% and 38 % in mice fed AAO and RNE diets, respectively. A reduction of cholesterol concentrations by 36 % and 38 % in animals receiving AAO and RNE diets was also observed *vs* STD diet (Table 4).

Diet	Glucose (mM)		Triglyceride (mM)		Cholesterol (mM)	
	Mean	SEM	Mean	SEM	Mean	SEM
STD	6.690[a]	0.25	3.070[a]	0.09	3.087[a]	0.10
RNE	5.932[b]	0.27	2.233[b]	0.27	1.876[b]	0.14
AAO	5.857[b]	0.31	2.378[b]	0.18	1.756[b]	0.09

(Mean values n=8 with their standard errors of the mean for each parameter measured) [a,b] Mean values with unlike superscript letters were significantly different (P≤0.05).

Table 4. Glucose, triglycerides and cholesterol levels in plasma of mice fed with a standard (STD) diet or diet supplemented with Raftiline (RNE) and *A. angustifolia* Oax. (AAO).

4. Discussion

The determination of the production of $C_{2:0}$, $C_{3:0}$ and $C_{4:0}$ acids in the caecum, proximal, medial and distal colon of mice fed with different diets, was performed with the aids to evaluate the profiles of these acids throughout the caecum, large intestine and faeces of mice, and also to be able to establish the main sites of fermentation of inulins such Raftiline (RNE) and fructans extracted from *A. Angustifolia* (AAO) supplemented in the diets. These objectives basically arise from the knowledge on the structural differences between the two fructans types. Raftiline is an inulin type fructan with an average DP of 25 and it is know that its structure is completely linear, therefore containing a terminal glucose molecule. On the other hand, *A. angustifolia* fructans have an average DP of 32, with molecular structures very complex, they are highly branched and present a terminal glucose molecule (graminans) or internal glucose (agavins) [11]. Many reports have established that the structure of undigested carbohydrates and the microbiota present in an ecosystem are determining factors that control fermentation in the gut [50]. It is also know that the profiles of the production and distribution of SCFAs in the gut are influenced not only by the type of consumed carbohydrates, but also by the place where fermentation of those carbohydrates takes place, essentially in the caecum in mice and in the upper colon in humans, but the type of substrate may also affect the site of fermentation [51]. Previous reports mentioned that high DP fructans are fermented more slowly in the caecum and proximal colon, thus reaching the distal colon almost unchanged [6, 34], but bacteria present in this section produce mainly $C_{4:0}$ acid as the end product of such fermentation [51]. In the present work, inulins were fermented poorly in the caecum and proximal colon but an increment was observed on the concentration of $C_{4:0}$ acid in the distal colon, which agrees well with previous reports. However, mice fed with agave fructans (AAO) produced greater amounts of SCFAs in the caecum and proximal and medial colon than those mice fed with inulins (RNE), suggesting that AAO fructans were easily fermentable, independently of the their high DP, therefore, the difference might be due to the presence of branches in the agave fructans, which could make the molecule more accessible to enzymes (fructosyltransferases), in other words, there are more terminal fructose available for the fructosyltransferases of bacteria. According to the production of SCFAs, proximal and medial colon, were the main sites where fermentation of *A. angustifolia* fructans was carried out. In general, an increment was observed in the production of $C_{3:0}$ and $C_{4:0}$ acids in the distal colon of mice that received diets supplemented with fructans. On the other hand, total SCFAs and individual SCFAs concentrations in the different sections of the colon are very important since they have been associated with many diseases of the colon, especially with colon cancer and gastrointestinal disorders. Therefore, increased SCFAs production and a greater delivery of them distally, especially $C_{4:0}$ acid, may have an important role in preventing some of these diseases and other metabolic problems. Moreover, $C_{3:0}$ acid has been reported to have a positive metabolic effect, through the inhibition of hepatic cholesterol synthesis from $C_{2:0}$ acid [32]. Interestingly, a significant decrease in plasma triglycerides and cholesterol levels of animals fed with fructans was observed in this work. Finally, the mice that consumed agave fructans showed the more pronounced drop on pH in the caecum and the three sections of the colon, which creates a more acid environment which is highly beneficial for the grow of bacteria such as *bifidobacteria* and/or *lactobacilli* but is detrimental for the growth of potentially pathogenic species [17,18].

As a general conclusion, we can mention that the supplementation of diets with inulins or agavins altered the large intestine environment by increasing the amounts of SCFAs and lowering the pH in the colon, consequently reducing few health risks. Finally, we would like to close this work saying that these SCFAs had a positive effect on the host lipid metabolism, since they decreased the levels of triglycerides, cholesterol and glucose in blood of mice fed with supplemented diets.

Based on all the previous data, agave fructans may offer a good prebiotic potential, opening new and excited alternatives as food supplements. Even do, further research is definitely needed on specific health problems and should be performed using supplemented diets with agavins of different structures as well as different mixtures and concentrations, because more knowledge is needed on health issues such obesity, diabetes, colon cancer and in general, gut associated risks that might be improved with this type of ingredients.

Author details

Alicia Huazano-García and Mercedes G. López

Departamento de Biotecnología y Bioquímica,
Centro de Investigación y de Estudios Avanzados del IPN, México

Metabolism of Short Chain Fatty Acids in the Colon and Faeces of Mice After a
Supplementation of Diets with Agave Fructans

149

Acknowledgement

Alicia Huazano-García thanks the Consejo Nacional de Ciencia y Tecnología (CONACYT) for her scholarship and also thanks M.S. Patricia Santiago for the agave fructans sample.

5. References

[1] Cairns A., Pollock C., Gallagher J., Harrison J. Fructans: Synthesis and Regulation. Leegood R., Sharkey T., Caemmerer S. von (eds) Photosynthesis: Physiology and Metabolism. Kluwer Academic Publishers; 2000. p 301-320.

[2] Niness K. Inulin and Oligofructose: What are They? The Journal of Nutrition 1999; 129(7) 1402S-1406.

[3] Van Loo J., Coussement P., De Leenheer L., Hoebregs H., Smits G. On the Presence of Inulin and Oligofructose as Natural Ingredients in the Western Diet. Critical Reviews in Food Science and Nutrition 1995; 35(6) 525-552.

[4] Ritsema T., Smeekens S. Engineering Fructan Metabolism in Plants. Journal of Plant Physiology 2003; 160(7) 811-820.

[5] Roberfroid M. Prebiotics: The Concept Revisited. The Journal of Nutrition 2007; 137(3) 830S-837.

[6] Roberfroid M., Van Loo J., Gibson G. The Bifidogenic Nature of Chicory Inulin and its Hydrolysis Products. The Journal of Nutrition 1998; 128(1) 11-19.

[7] Good-Avila S., Souza V., Gaut B., Eguiarte L. Timing and Rate of Speciation in Agave (Agavaceae). Proceedings of the National Academy of Sciences of the United States of America 2006;103(24) 9124-9129.

[8] Eguiarte L., Souza V., Silva-Montellano A. Evolución de la Familia Agavaceae: Filogenia, Biología Reproductiva y Genética de Poblaciones. Bolletin de la Sociedad Botánica de México 2000; 66 131-150.

[9] Gentry H. Agaves of continental North America. The University of Arizona Press; 1998.

[10] López M., Mancilla-Margalli N., Mendoza-Díaz G. Molecular Structures of Fructans from Agave tequilana Weber var. azul. Journal of Agricultural and Food Chemistry 2003; 51(27) 7835-7840.

[11] Mancilla-Margalli N., López G. Water-soluble Carbohydrates and Fructan Structure Patterns from Agave and Dasylirion Species. Journal of Agricultural and Food Chemistry 2006; 54(20) 7832-7839.

[12] Roberfroid M. Introducing Inulin Type Fructans. British Journal of Nutrition 2005;93(1) 13S-25.

[13] Roberfroid M. Prebiotics and Probiotics: Are they Functional Foods? The American Journal of Clinical Nutrition 2000; 71(1) 1682S-1687.

[14] Kolida S., Gibson G. Prebiotic Capacity of Inulin-type Fructans. The Journal of Nutrition 2007; 137(11) 2503S-2506.

[15] López M., Urías-Silvas J. Agave Fructans as Prebiotics. Norio S., Noureddine B., Shuichi O. (eds) Recent Advances in Fructooligosaccharides Research. Research Signpost; 2007. p 1-14.

[16] Gibson G., Probert H., Van Loo J., Rastall R., Roberfroid M. Dietary Modulation of the Human Colonic Microbiota: Updating the Concept of Prebiotics. Nutrition Research Reviews 2004; 17(2) 259-275.

[17] Gibson G., Roberfroid M. Dietary modulation of the human colonic microbiota: introducing the concept of prebiotics. The Journal of Nutrition 1995; 125(6) 1401-1412.

[18] Roberfroid M. Fructo-oligosaccharide Malabsorption: Benefit for Gastrointestinal Functions. Current Opinion in Gastroenterology 2000; 16(2) 173-177.

[19] Topping D., Clifton P. Short-chain Fatty Acids and Human Colonic Function: Roles of Resistant Starch and Nonstarch Polysaccharides. Physiological Reviews 2001; 81(3) 1031-1063.

[20] Gibson G. Dietary Modulation of the Human Gut Microflora Using the Prebiotics Oligofructose and Inulin. The Journal of Nutrition 1999; 129(7) 1438S-1441.

[21] Cummings J., Macfarlane G. The Control and Consequences of Bacterial Fermentation in the Human Colon. Journal of Applied Bacteriology 1991; 70(6) 443-459.

[22] Bernalier A., Dore J., Durand M. Biochemistry of fermentation. Gibson G., Roberfroid M. (eds) Colonic Microbiota, Nutrition and Health. Dordrecht:Kluwer Academic Publishers; 1999. p 37-53.

[23] Flamm G., Glinsmann W., Kritchevsky D., Prosky L., Roberfroid M. Inulin and Oligofructose as Dietary Fiber: A Review of Evidence. Critical Reviews in Food Science and Nutrition 2001; 41(5) 353;362.

[24] Wong J., De Souza R., Kendall C., Emam A., Jenkins D. Colonic Health: Fermentation and Short Chain Fatty Acids. Journal of Clinical Gastroenterology 2006; 40(3) 235-243.

[25] Cummings J., Roberfroid M., Andersson H., Barth C., Ferro-Luzzi A., Ghoos Y., Gibney M., Hermonsen K., James W., Korver O. A New Look at Dietary Carbohydrate: Chemistry, Physiology and Health. European Journal of Clinical Nutrition 1997; 51(7) 417-423.

[26] Cook S., Sellin J. Short Chain Fatty Acids in Health and Disease. Alimentary Pharmacology and Therapeutics 1998; 12(6) 499-507.

[27] Macfarlane G., Macfarlane S. Models for Intestinal Fermentation: Association between Food Components, Delivery Systems, Bioavailability and Functional Interactions in the Gut. Current Opinion in Biotechnology 2007; 18(2) 156-162.

[28] Zeng J., Tan Z. Metabolic Homeostasis and Colonic Health: The Critical Role of Short Chain Fatty Acids. Current Nutrition and Food Science 2010; 6(3) 209-222.

[29] Pomare E., Branch W., Cummings J. Carbohydrate Fermentation in the Human Colon and its Relation to Acetate Concentrations in Venous Blood. The Journal of Clinical Investigation 1985; 75(5) 1448-1454.

[30] Cummings J. Short Chain Fatty Acids in the Human Colon. GUT 1981; 22(9) 763-779.

[31] Venter C., Vorster H., Cummings J. Effects of Dietary Propionate on Carbohydrate and Lipid Metabolism in healthy volunteers. The American Journal of Gastroenterology 1990; 85(5) 549-553.

[32] Cheng H., Lai M. Fermentation of resistant rice starch produces propionate reducing serum and hepatic Cholesterol in Rats. The Journal of Nutrition 2000; 130(8) 1991-1995.

[33] Pryde S., Duncan S., Hold G., Stewart C., Flint H. The Microbiology of Butyrate Formation in the Human Colon. FEMS Microbiology Letters 2002; 217(2) 133-139.

[34] Van de Wiele T., Boon N., Possemiers S., Jacobs H., Verstraete W. Inulin-type Fructans of Longer Degree of Polymerization Exert More Pronounced *in vitro* Prebiotic Effects. Journal of Applied Microbiology 2007; 102(2) 452-460.

[35] Stewart M., Timm D., Slavin J. Fructooligosaccharides Exhibit More Rapid Fermentation than Long-chain Inulin *in* an *vitro* fermentation system. Nutrition Research 2008; 28(5) 329-334.

[36] Rycroft C., Jones M., Gibson G., Rastall R. A Comparative *in vitro* Evaluation of the Fermentation Properties of Prebiotic Oligosaccharides. Journal of Applied Microbiology 2001; 91(5) 878-887.

[37] Urías-Silvas J., López M. *Agave* spp. and *Dasylirion* sp. Fructans as a Potential Novel Source of Prebiotics. Dynamic Biochemistry, Process Biotechnology and Molecular Biology 2009; 3(1) 59-64.

[38] Santiago-García P., López M. Prebiotic Effect of Agave Fructans and Mixtures of Different Degrees of Polymerization from *Agave angustifolia* Haw. Dynamic Biochemistry, Process Biotechnology and Molecular Biology 2009; 3(1) 52-58.

[39] Gomez E., Tuohy K., Gibson G., Klinder A., Costabile A. *In vitro* Evaluation of the Fermentation Properties and Potential Prebiotic Activity of Agave Fructans. Journal of Applied Microbiology 2010; 108(6) 1859-2228.

[40] Nilsson U., Nyman M. Short-chain Fatty Acid Formation in the Hindgut of Rats Fed Oligosaccharides Varying in Monomeric Composition, Degree of Polymerization and Solubility. British Journal of Nutrition 2005; 94(5) 705-713.

[41] Licht T., Hansen M., Poulsen M.,Dragsted L. Dietary Carbohydrate Source Influences Molecular Fingerprints of the Rat Faecal Microbiota. BMC Microbiology 2006; 6(1) 98-107.

[42] Klessen B., Hartmann L., Blaut M. Oligofructose and Long-chain Inulin: Influence on the Gut Microbial Ecology of Rats Associated with a Human Faecal Flora. British Journal of Nutrition 2001; 86(2) 291-300.

[43] Levrat M., Remesy C., Demigne C. High Propionic Acid Fermentations and Mineral Accumulation in the Cecum of Rats Adapted to Different Levels of Inulin. The Journal of Nutrition 1991; 121(11) 1730-1737.

[44] Levrat M., Favier M., Moundras C., Remesy C., Demigne C., Morand C. Role of Dietary Propionic Acid and Bile Acid Excretion in the Hypocholesterolemic Effects of Oligosaccharides in Rats. The Journal of Nutrition 1994; 124(4) 531-538

[45] Daubioul C., Rousseau N., Demeure R., Gallez B., Taper H., Declerck B., Delzenne N. Dietary Fructans, but not Cellulose, Decrease Triglyceride Accumulation in the Liver of Obese Zucker *fa/fa* Rats. The Journal of Nutrition 2002; 132(5) 967-973.

[46] Trautwein E., Rieckhoff D., Erbersdobler H. Dietary Inulin Lower Plasma Cholesterol and Triacylglycerol and Alters Biliary Bile Acid Profile in Hamsters. The Journal of Nutrition 1998; 128(11) 1937-1943.

[47] Urías-Silvas J., Cani P., Delmee E., Neyrinck A., López M., Delzenne N. Physiological Effects of Dietary Fructans Extracted from *Agave tequilana* Gto. and *Dasylirion* spp. British Journal of Nutrition 2008; 99(2) 254-261.

[48] García-Pérez M. Efecto de los Fructanos de *Dasylirion* sp. en la Secreción de Grelina y GLP-1 en Ratones. MS thesis. CINVESTAV-Irapuato; 2008.

[49] Pietro F., Luceri C., Dolara P., Giannini A., Biggeri A., Salvadori M., Clune Y., Collins K., Paglierani M., Cademi G. Antitumorigenic Activity of the Inulin Enriched with Oligofructose in Combination with the Probiotics *Lactobacillus rhamnosus* and *Bifidobacterium lactis* on Azoximethane-induced Colon Carcinogenesis in Rats. Carcinogenesis 2002; 23(11) 1953-1960.

[50] Henningsson A., Björck I., Nyman G. Combinations of Indigestible Carbohydrates Affect Short-chain Fatty Acid Formation in the Hindgut of Rats. The Journal of Nutrition 2002; 132(10) 3098-3104.

[51] Hughes R., Rowland I. Stimulation of Apoptosis by Two Prebiotic Chicory Fructans in the Rat Colon. Carcinogenesis 2000; 22(1) 43-47.

Chapter 8

Metabolism of Plasma Membrane Lipids in Mycobacteria and Corynebacteria

Paul K. Crellin, Chu-Yuan Luo and Yasu S. Morita

Additional information is available at the end of the chapter

1. Introduction

Bacteria of the Corynebacterineae, a suborder of the Actinobacteria, comprise *Mycobacterium, Corynebacterium, Nocardia, Rhodococcus* and other genera. This suborder of high GC gram-positive bacteria includes a number of important human pathogens, such as *Mycobacterium tuberculosis, Mycobacterium leprae* and *Corynebacterium diphtheriae*, the causative agents of tuberculosis, leprosy and diphtheria, respectively. *M. tuberculosis* is the most medically significant species, a devastating human pathogen infecting around one-third of the entire human population and responsible for more than 1 million deaths annually. The Corynebacterineae also includes non-pathogenic species such as *Mycobacterium smegmatis*, a saprophytic species, and *Corynebacterium glutamicum*, an industrial workhorse for the production of amino acids and other useful compounds. These relatively fast-growing species serve as useful models to study metabolic processes essential to the growth and survival of the slow-growing pathogens.

All these bacteria share a common feature, a distinctive multilaminate cell wall composed of peptidoglycan, complex polysaccharides, and both covalently linked lipids and free lipids/lipoglycans (Fig. 1). Among them, mycolic acids are the hallmark of these species. These long chain α-branched, β-hydroxylated fatty acids are covalently linked to the arabinogalactan polysaccharide layer. This mycolic acid layer is complemented by a glycolipid layer to form an outer "mycomembrane" analogous to the outer membrane of Gram-negative bacteria. [1, 2]. The outer leaflet of the mycomembrane is composed of a variety of lipids including trehalose dimycolates (TDMs), glycopeptidolipids (GPLs), phthiocerol dimycocerosates (PDIMs), sulfolipids, phenolic glycolipids (PGLs), and lipooligosaccharides. Some of these lipids are widely distributed while others are restricted to particular species. For example, TDMs and their structural equivalents are found in both mycobacteria and corynebacteria, while PDIMs and PGLs are restricted to a subset of mycobacteria. The structure and hydrophobic properties of the cell wall make it a potent permeability barrier that is responsible for intrinsic resistance of mycobacteria to an array of host microbiocidal processes, many antibiotics and sterilization conditions [3, 4]. Many of the cell wall components of pathogenic mycobacterial species are essential for pathogenesis and *in vitro* growth, hampering efforts to characterize the function of individual proteins in their assembly. In contrast, some non-pathogenic species such as *C. glutamicum* can tolerate the loss of major cell wall components, making them useful model systems for delineating processes involved in the assembly of core cell wall structures.

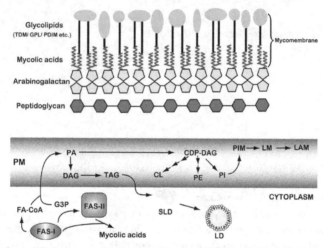

Figure 1. Mycobacterial plasma membrane and cell wall with flow of key metabolic pathways. Some of the metabolites are exported to the mycomembrane. SLD, small lipid droplet; LD, lipid droplet; FA-CoA, fatty acyl-CoA. See text for other abbreviations used in the figure.

Studies on mycobacteria and corynebacteria provide a unique opportunity to illustrate the complexity and diversity of lipid metabolic pathways in bacteria. They have a significantly higher lipid content than other bacteria with cell wall lipids comprising ~40% of the dry cell mass. *M. tuberculosis* produces a diversity of lipids unparalleled in bacteria, from simple fatty acids to highly complex long chain structures such as mycolic acids. It has devoted a significant proportion of its coding capacity to lipid metabolism and produces about 250 enzymes dedicated to fatty acid metabolism, which is around five times the number produced by *Escherichia coli* [5]. Lipid biosynthesis places a significant metabolic burden on the organism but is ultimately advantageous, allowing *M. tuberculosis* to survive and replicate in the inhospitable environment of host macrophages. While capable of *de novo* synthesis, these bacteria also scavenge and degrade host cell membrane lipids to acetyl-CoA, via broad families of β-oxidation and other catabolic enzymes, for incorporation into their own metabolic pathways and to fuel cellular processes.

The plasma membrane provides the platform for lipid metabolism. While some lipid metabolic reactions take place in the cytoplasm or cell wall, the plasma membrane is the pivotal site for the metabolism of lipids. At the same time, this membrane must perform many other functions associated with energy production, nutrient uptake, protein export, and various sensing/signaling reactions. Studies on how these metabolic and cellular processes might be organized within bacterial plasma membranes are in their infancy. Understanding the homeostasis of the plasma membrane is particularly important in Corynebacterineae organisms because this structure must support the high biosynthetic demands of sustaining such a lipid-rich cell wall. In this chapter, we focus our discussion on processes of lipid metabolism that are critical for the biogenesis and maintenance of the plasma membrane, and illustrate the recent progress on our understanding of plasma membrane biogenesis in mycobacteria and corynebacteria.

2. Functions of plasma membrane lipids in mycobacteria and corynebacteria

In this section we will describe the functions of plasma membrane lipids. First, we will describe the functions of major structural phospholipids. We will then describe quantitatively minor lipids, which have important metabolic/physiological functions. Lastly, we will discuss the functions of neutral lipids because their biosynthesis is closely linked to phospholipid metabolism and neutral lipid storage is a critical part of plasma membrane homeostasis.

2.1. Structural lipids

Major structural components of the mycobacterial plasma membrane are phospholipids such as cardiolipin (CL), phosphatidylethanolamine (PE), phosphatidylinositol (PI), and glycosylated PIs (*i.e.* phosphatidylinositol mannosides (PIM), lipomannans (LM) and lipoarabinomannans (LAM), see below). The ratio of these phospholipids may vary depending on the species and growth conditions [6-8]. For example, one study indicated that CL, PE, and PI/PIMs represent about 37, 32, and 28%, respectively, of the total phospholipids in the plasma membrane in *M. smegmatis* [9], while another reported the ratio in *Mycobacterium phlei* to be about 50, 10, and 40% [10]. Phosphatidylglycerol (PG), which is abundant in other bacteria, is a relatively minor species in mycobacteria. Deletion of the PI biosynthetic gene has been shown to be lethal in *M. smegmatis* [9], indicating that PI or glycosylated PIs are essential for mycobacterial viability. In *M. tuberculosis*, putative PI synthetase (*Rv2612c*) and PGP synthetase (*Rv2746c*, involved in CL synthesis) genes are predicted to be essential [11], while the PS synthetase gene (*Rv0436c*, involved in PE synthesis) is not [12]. In corynebacteria, major species of phospholipids are PI, PG, CL, and acylphosphatidylglycerol (APG) [13], and PE appears to be absent.

CL is widely found in both prokaryotes and eukaryotes. It forms aggregates within the membrane bilayer. Nonyl acridine orange (NAO) is a fluorescent dye which is proposed to bind the hydrophobic surface created by the CL cluster [14], allowing microscopic visualization of CL domains. Indeed, using NAO, CLs were found to be enriched in septa and poles of actively dividing *M. tuberculosis* and *M. smegmatis* cells [15, 16]. CL has a non-bilayer structure [17, 18], and carries a small partially immobilized head group that is more exposed to the aqueous environment than those of other glycerophospholipids [19]. Although the physiological function of CL is unclear, its physical properties may indicate that it provides a platform for membrane-protein interactions. Indeed, some mycobacterial enzymes require CL for activity [20-22], although the molecular basis for these observations has not been clarified. Recent fractionation studies in *C. glutamicum* revealed that CL (as well as other phospholipids) is enriched in the plasma membrane [23, 24]. However, a large proportion of CL is also found to be associated with the outer membrane [24], suggesting that some of these phospholipids are exported to the outer membrane in corynebacteria. Similarly, CL is released from *M. bovis* bacillus Calmette-Guerin residing in host phagosomes, and converted to lyso-CL by a host phospholipase A_2 [25]. It has been suggested that lyso-CL may influence host immune responses during infection.

PE is another major class of glycerophospholipids in mycobacteria. Although PE is generally found in all organisms, it is particularly abundant in bacterial plasma membranes [26]. Mycobacteria are no exception [20], but corynebacteria apparently lack the capacity to synthesize PE [27]. Indeed, PE biosynthetic enzymes, such as PS synthetase and PS decarboxylase, appear to be absent in corynebacterial genomes. *Corynebacterium aquaticum* has been reported to possess PE [28], but this species was later reclassified as *Leifsonia aquatica* [29], which belongs to the suborder Micrococcineae of the order Actinomycetales. The functions of PE remain elusive at the molecular level, but it appears to play important roles as a component of the plasma membrane. For example, TBsmr, a small multidrug resistance family protein from *M. tuberculosis*, shows enhanced catalytic activities when PE is supplemented in a reconstituted liposome [30].

PIs are an important class of phospholipids, and are known to be further modified by extensive glycosylation. The resultant lipoglycans, termed PIMs, LM, and LAM, are essential structural components of mycobacterial and corynebacterial cell walls. Furthermore, in pathogenic species, they have been suggested to perform additional roles in the modulation of host immune responses in favor of the pathogen through myriad effects on macrophages including cytokine production, inhibition of phagosome maturation and apoptosis [31-34]. PIMs are oligo-mannosylated PIs carrying up to 6 mannose residues while LM/LAM carry much longer mannose polymers with arabinan modifications. It remains controversial if these glycolipids are embedded in the plasma membrane or exported to the outer membrane. A recent study suggests that LM/LAM appear to be anchored to both the plasma membrane and outer membrane [35]. In *C. glutamicum*, the outer membrane and plasma membrane were fractionated on sucrose gradients upon cell lysis, and the analysis of these membrane sub-fractions demonstrated that PIMs, LM and LAM are all enriched in the plasma membrane fraction [23]. Another recent study also suggested that PI/PIMs are major components of the plasma membrane of *C. glutamicum* [24]. In the latter study, however, substantial amounts of PI/PIMs were detected in the outer membrane as well. The functional significance of these subcellular localizations, as well as the physiological roles of LM/LAM in each of these locations, remain important questions. The structural importance of PIMs remains unclear as well. For example, a *pimE*-deletion mutant that cannot produce mature

PIM6 species (see below) is viable, but shows severe plasma membrane abnormalities [36], suggesting that higher order PIMs may be involved in the maintenance of plasma membrane integrity.

It is notable that some unusual phospholipids have been identified in corynebacteria. APG is an acylated form of PG which is widespread in corynebacteria [37-40], and is a major phospholipid species in *Corynebacterium amycolatum*. Another interesting phospholipid from *C. amycolatum* is acyl-phosphatidylinositol (API), which was identified by electrospray ionization mass spectroscopy [41]. *C. amycolatum* lacks a mycolic acid-based outer membrane, and does not appear to have a fracture plane other than the plasma membrane [42]. Therefore, APG and API are likely to be components of the plasma membrane, and are suggested to play structural roles. Very little is known about their biosynthesis, and acyltransferases responsible for their synthesis remain to be identified for both lipid species.

2.2. Functional lipids

There are some examples of lipids that appear to play no structural roles in the plasma membrane. They often exist in low quantities but play important functional roles. Among these, polyprenol-phospho-sugars function as sugar donors. Two well-studied examples are polyprenol phosphomannose (PPM) and decaprenol phosphoarabinose (DPA). These molecules are the donors of mannose and arabinose, respectively, and their biosynthesis will be discussed in a later section.

PI 3-phosphate, recently identified in both *M. smegmatis* and *C. glutamicum* [43], may prove to be another interesting example of a functional lipid. It accumulates only transiently upon stimulation by high concentrations of salt, and behaves as if it is involved in a signaling cascade. However, whether PI 3-phosphate represents a mediator of stress responses remains to be addressed. More recently, lysinylated PG was identified as a minor phospholipid species in *M. tuberculosis* [44]. The synthesis of lysinylated PG is mediated by LysX and a *lysX* deletion mutant showed altered phospholipid metabolism and membrane integrity [16, 44], suggesting a regulatory role of lysinylated PG in plasma membrane homeostasis.

Carotenoids are photo-protective pigments and serve to scavenge free radicals or harvest light [45]. Several mycobacterial species are known to produce carotenoids with the notable exception of *M. tuberculosis*, despite the presence of a carotenoid oxidase in the human pathogen [46]. These hydrophobic pigments are thought to be present in the plasma membrane but whether they play structural roles in addition to a photo-protective role remains to be elucidated.

2.3. Lipid storage for energy and carbon

Neutral lipids are an important reservoir of stored energy and carbon, and their metabolism is closely linked to plasma membrane phospholipid metabolism. Unlike many other bacteria which use polyhydroxyalkanoates as a lipid storage material [47], Actinobacteria use triacylglycerides (TAGs) as a major form of lipid storage, and the presence of TAGs has been reported in *Mycobacterium*, *Streptomyces*, *Rhodococcus* and *Nocardia* [48-52]. Interestingly, corynebacteria seem to lack the capacity to synthesize TAG, indicating that some lineages of Actinobacteria have eliminated this capacity at some point in their evolution. Recent evidence suggests that *M. tuberculosis* accumulates TAG-based lipid droplets while residing in macrophages using fatty acids released from host TAGs, and this process is critical for acquiring a dormancy phenotype [53]. Nevertheless, a mutant defective in accumulating TAG remained viable under *in vitro* dormancy-inducing conditions [54]. These somewhat contradictory observations suggest that our understanding of TAG metabolism in mycobacteria is far from complete. As we illustrate later, there appear to be several redundant genes involved in the final step of TAG synthesis, suggesting that it is an important regulatory step of lipid metabolism in these bacteria.

Cholesterol has recently been suggested to be an alternative form of lipid storage in mycobacteria. Neither mycobacteria nor corynebacteria have the capacity to synthesize cholesterol. However, cholesterol is taken up by *M. tuberculosis* cells residing in the host, and components of the *mce4* operon have been shown to be involved in cholesterol import [55]. Cholesterol catabolism is critical in the chronic phase of animal infection, and a fully functional catabolic pathway is encoded by the *M. tuberculosis* genome [56]. Furthermore, cholesterol appears to accumulate in the mycobacterial cell envelope, and this might represent a potential form of lipid storage for *M. tuberculosis* during animal infection [57, 58]. Although the authors of this study suggested that cholesterol accumulates in the outer membrane, it remains possible that the plasma membrane is the true site of accumulation. Therefore, in addition to acting as a lipid storage molecule, cholesterol may play roles in plasma membrane structure and function, and these possibilities await further exploration.

Catabolism of cholesterol, amino acids and odd-chain-length/methyl branched fatty acids produces propionyl-coenzyme A (CoA). Propionate accumulation has been shown to be toxic in various organisms [59-61], and *M. tuberculosis* has multiple pathways to metabolize propionyl-CoA [62]. Metabolized propionyl-CoA is in part incorporated into TAG [63], and it has been suggested that TAG functions as a sink for reducing equivalents in addition to being a source of carbon and energy.

3. Structure and metabolism of plasma membrane lipids in mycobacteria and corynebacteria

In this section, we will describe the structure and metabolism of various lipids found in the plasma membrane of mycobacteria and corynebacteria in more detail. Lipids are categorized into the following four classes based on their key structural features.

3.1. Fatty acids

M. tuberculosis devotes a large proportion of its coding capacity to genes involved in fatty acid metabolism [5], highlighting the importance of lipids to the organism. Fatty acid metabolism is essential for intracellular survival of the pathogen since it forms the precursors of key membrane components such as plasma membrane phospholipids and outer membrane glycolipids. In particular, mycolic acids, which are very long chain α-alkyl β-hydroxy fatty acids, form the hydrophobic, protective mycomembrane described earlier. *M. tuberculosis* encodes two distinct enzyme systems for biosynthesis of fatty acids, designated FAS (fatty acid synthase) I and II (Fig. 2). Studies on fatty acid synthesis date back to the 1970s when *M. smegmatis* was shown to contain both type I fatty acid synthetase (FAS-I), involving a large multifunctional polypeptide, and type II fatty acid synthetase (FAS-II), consisting of a series of distinct enzymes [64]. The key elongation unit is malonyl-CoA, which is produced by acetyl-CoA carboxylase (ACCase) and the *M. tuberculosis* genome encodes several such enzymes (AccA1-3 and AccD1-6). The resultant malonyl-CoA is incorporated into fatty acids by the two FAS systems.

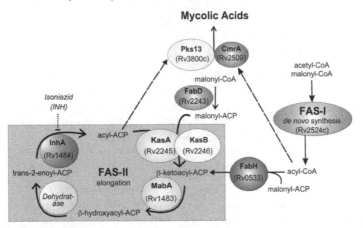

Figure 2. Fatty acid biosynthesis pathways in mycobacteria. Point of inhibition by the front-line tuberculosis drug isoniazid is indicated. Product profile of FAS-I is bimodal, and C_{16}-C_{18}-CoA and C_{24}-C_{26}-CoA are produced. Dashed lines indicate that some of the fatty acid products are further utilized for mycolic acid production.

3.1.1. De novo synthesis by FAS-I

Surprisingly, members of the Corynebacterineae use a eukaryote-like FAS-I system for *de novo* fatty acid synthesis. The single, essential [11], 9.2kb *fas* gene encodes a 326 kDa protein containing all seven domains necessary to perform the iterative series of reactions: acyl transferase, enoyl reductase, β-hydroxyacyl dehydratase, malonyl transferase, acyl carrier protein, β-ketoacyl reductase, and β-ketoacyl synthase [65, 66]. This very large protein elongates acetyl groups by 2-carbon (acetate) units using acetyl-CoA and malonyl-CoA. Early rounds of elongation yield C_{16} to C_{18}-CoA products that are used for synthesis of membrane phospholipids or to feed into the FAS-II system. More extensive elongation yields C_{24}-C_{26} products that ultimately form the α-branch of mycolic acids. Unlike *M. tuberculosis*, *C. glutamicum* encodes two *fas* genes (*fasA* and *fasB*) with FasA taking the dominant role [67]. The presence of two Fas proteins may compensate for the lack of a FAS-II system in this organism.

3.1.2. Elongation by FAS-II

The FAS-II system is commonly found in bacteria and plants and, unlike FAS-I, is composed of a series of separate enzymes, each performing one step in the pathway. FAS-II elongates medium chain fatty acids derived from FAS-I using malonyl-CoA, producing C_{18}-C_{30} fatty acids [68]. FAS-II has been extensively studied in E. coli [69] and orthologs of the *fab* genes have been identified in mycobacteria. AcpM is a mycobacterial acyl carrier protein (ACP) and plays a key role in transferring acyl groups between the various enzyme components [70]. The seven genes are located in two clusters on the M. tuberculosis chromosome [5], comprising *mtfabD-acpM-kasA-kasB-accD6* and *mabA-inhA*. Initially, the malonate group is transferred from malonyl-CoA to AcpM by the MtFabD protein. Then MtFabH performs a Claisen condensation of malonyl-ACP with acyl-CoA to form β–ketoacyl-ACP. A four-step cycle is then initiated [64] in which:

1. β–ketoacyl-ACP reductase MabA reduces the β–keto group with concomitant oxidation of NADPH
2. β-hydroxyacyl-ACP dehydratase dehydrates the β-hydroxyl to enoyl-ACP
3. enoyl-ACP reductase InhA, a target of the first-line anti-tuberculosis drug isoniazid (INH) [71], reduces enoyl-ACP to acyl-ACP with concomitant oxidation of NADPH
4. β-ketoacyl-ACP synthase KasA/B elongates acyl-ACP by 2 carbon units, forming β-ketoacyl-ACP, which can feed back into step 1.

In this way, the hydrocarbon chain increases by 2 carbons each cycle. Further elongation and processing of the products of FAS-II produces the precursors of the long meromycolate chains that are condensed with the α-branches derived from FAS-I by the large polyketide synthase Pks13 [72]. Reduction of the β–keto group by CmrA forms the mature C_{60}-C_{90} mycolic acid [73].

3.2. Glycerolipids

Glycerolipids include both nonpolar lipids and polar phospholipids. Their biosynthesis is overlapping and 1,2-diacyl-*sn*-glycerol 3-phosphate, commonly known as phosphatidic acid (PA), is an important intermediate at the branch point (Fig. 3) [74]. In this section, we focus our discussion on the biosynthesis of PA and its conversion to non-polar lipids. Non-polar lipids are generally divided into three different classes depending on the number of fatty acids attached to glycerol: monoacylglycerol (MAG), diacylglycerol (DAG) and TAG. TAG is a glycerol carrying three fatty acyl chains, and its biosynthesis diverges from phospholipid synthesis after the synthesis of PA. TAG is a major component of lipid droplets, which accumulate in the cytoplasm. How TAG is made in the plasma membrane and incorporated into lipid droplets remains largely unclear. Here, we provide an overview of the TAG metabolic pathway.

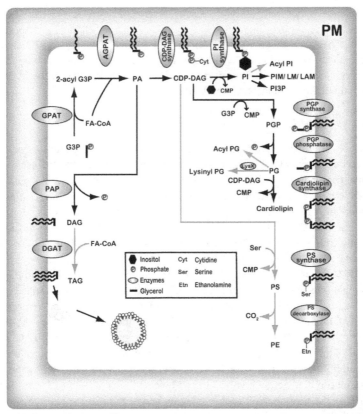

Figure 3. Glycerolipid/phospholipid biosynthesis pathways. Some pathways such as TAG and PE biosynthesis (shown as green arrows) do not occur in corynebacteria while some others (shown as blue arrows) are known to occur only in corynebacteria. PG is abundant in corynebacteria, but is a minor species in mycobacteria.

3.2.1. Biosynthesis of PA

The first step of PA biosynthesis is mediated by glycerol phosphate acyltransferase (GPAT) transferring an acyl chain from acyl-CoA to glycerol-3-phosphate, forming acyl-glycerol 3-phosphate. In general, this reaction produces 1-acyl-*sn*-glycerol 3-phosphate. However, mycobacteria are unusual in that 2-acyl-*sn*-glycerol 3-phosphate is used as the main intermediate for the production of PA [75]. Another unusual feature is that oleic acid, an unsaturated fatty acid often found at the *sn*-2 positions of glycerolipids, is found at the *sn*-1 position in mycobacteria. Instead, palmitic acid, a saturated fatty acid, is the preferred fatty acid attached to the *sn*-2 position in mycobacteria [75, 76]. In the second step, acylglycerol phosphate acyltransferase (AGPAT) further transfers a fatty acid from acyl-CoA to 2-acyl-*sn*-glycerol 3-phosphate, producing PA. PA can be diverted to TAG synthesis, or activated to form cytidine diphosphate-diacylglycerol (CDP-DAG), which is the precursor for the synthesis of phospholipids. Therefore, PA represents an important branch point for the synthesis of TAG and phospholipids [74]. An alternative pathway for PA synthesis is phosphorylation of DAG by DAG kinase, and Rv2252 has been suggested to be involved in this reaction [77]. Disruption of this enzyme results in altered PIM biosynthesis, but precise functions of this metabolic pathway remain unclear.

3.2.2. TAG Biosynthesis

TAG is *de novo* synthesized by two steps. First, PA is dephosphorylated to become DAG, and this reaction is mediated by phosphatidic acid phosphatase (PAP). PAP was discovered from animal tissues in 1957 by the group of Eugene Kennedy [78], and the gene encoding this activity was recently identified in *Saccharomyces cerevisiae* [79]. Nothing is known about this enzyme in mycobacteria or corynebacteria. In the second step, diacylglycerol acyltransferase (DGAT) catalyzes the addition of a fatty acyl-CoA to DAG to form TAG. Until recently, little was known about the genes involved in this final step of TAG synthesis in mycobacteria. Analysis of this final step is complicated because there are multiple genes encoding TAG synthetase in mycobacteria and corynebacteria. For example, the *M. tuberculosis* genome encodes 15 putative TAG synthetase genes [48, 80]. Despite the redundancies, recent studies reported that some of these *tgs* genes are critical for TAG synthesis in *M. tuberculosis* [48, 54]. Specifically, TAG synthetases encoded by *Rv3130c* (*tgs1*), *Rv3734c* (*tgs2*), *Rv3234c* (*tgs3*), and *Rv3088* (*tgs4*) have been shown to have TAG synthetase activities [53]. Furthermore, Tgs1 has been demonstrated to be the main contributor to TAG synthesis and lipid droplet formation in *M. tuberculosis* [53]. More recently, Ag85A, which is known as a mycolyltransferase involved in TDM biosynthesis, was shown to possess DGAT activity [81]. Ag85A is not homologous to other *tgs* genes, and may represent a novel class of TAG biosynthetic enzymes. TAG not only forms a lipid droplet in the cytoplasm, but also accumulates in the cell wall of mycobacteria [82]. Therefore, Ag85A located in the cell wall might be involved in the production of surface-exposed TAGs.

3.2.3. Utilization of TAG

Under starvation conditions where stored TAG needs to be mobilized for energy production, TAG is catabolized by lipases. In 1977, TAG lipase was purified from stationary phase *M. phlei* and predicted to have a molecular weight of about 40 kDa [83]. More recently, LipY, encoded by the *M. tuberculosis Rv3097c* gene, was identified as a TAG lipase [84]. LipY appears to play a critical role in TAG catabolism because a *M. tuberculosis lipY* deletion mutant cannot utilize accumulated TAG under starvation conditions. Another recent study demonstrated that LipY has a dual localization pattern [85]: while a fraction of LipY was found in the cytoplasm, consistent with its role in the catabolism of intracellular TAG, a significant fraction of LipY was also localized to the outer membrane of the cell wall, indicating that it may be involved in the breakdown of exogenously available TAGs. Indeed, it has been long known that *M. tuberculosis* depends on fatty acids as a preferred energy source during infection [86], and LipY may well be a critical enzyme for the utilization of host lipids during an *M. tuberculosis* infection. Another lipase encoded by *Rv0183* shows preference for MAG over DAG and TAG, and is localized to the cell wall [87], suggesting its involvement in subsequent reactions of TAG breakdown. However, whether it is involved in degradation of host-derived TAG or intracellular TAG remains to be determined.

3.2.4. Lipid droplet formation

In eukaryotes, lipid droplets form in between the two leaflets of the endoplasmic reticulum membrane [88]. In bacteria, a distinct mechanism of lipid body formation has been proposed. For example, in rhodococci, TAG is formed in the cytoplasmic surface of the plasma membrane. Small lipid droplets are then fused to each other, coated by a monolayer of phospholipids, and released from the surface of the plasma membrane into the cytoplasm as mature lipid droplets [89]. Although no endogenous proteins have been found to associate with lipid droplets in rhodococci or mycobacteria, heterologous expression of known lipid droplet-associated proteins resulted in correct targeting of these proteins to lipid droplets in both *R. opacus* and *M. smegmatis* [90, 91], allowing visualization of lipid droplets in these organisms.

3.3. Phospholipids

3.3.1. CDP-DAG

In both eukaryotic and prokaryotic cells, PA is activated by CTP to form CDP-DAG, and this reaction is mediated by CDP-DAG synthase [92]. The synthesis of CDP-DAG commits the pathway to phospholipid biosynthesis, and CDP-DAG is a common precursor for the biosynthesis of all glycerophospholipids in mycobacteria and corynebacteria. The activity of CDP-DAG synthetase is associated with plasma membrane in *M. smegmatis*, and is possibly encoded by the *cdsA* (*Rv2881c*) gene in *M. tuberculosis* H37Rv [93].

3.3.2. CL

CL is composed of four acyl chains, three glycerols and two phosphates, and is structured in a 1,3-diphosphatidylglycerol configuration [94]. It is a common phospholipid in bacteria, and is one of the abundant phospholipids in mycobacteria and corynebacteria. To initiate CL synthesis, PG phosphate synthase first produces PG phosphate (PGP) using CDP-DAG and glycerol 3-phosphate as substrates. An *M. smegmatis* strain engineered to overexpress *M. tuberculosis* PgsA3 (encoded by *Rv2746c*) was shown to overproduce PG, suggesting that PgsA3 is the PGP synthase [9]. PGP is then converted into PG via PGP phosphatase. Three phosphatases, PgpA, PgpB, and PgpC, have been identified as PGP phosphatases in *E. coli* [95-97]. Furthermore, Gep4 and PTPMT1 have been identified as PGP phosphatases in yeast and mammals, respectively [98, 99]. Some homologs exist in the genomes of mycobacteria and corynebacteria, but experimental verification of these genes remains to be performed.

Typically, the final step of CL synthesis in prokaryotes is mediated by a reaction that utilizes two PG molecules, producing one molecule of CL and one molecule of glycerol. However, in mycobacteria, the eukaryote-like reaction, which utilizes PG and CDP-DAG to produce CL, has been shown to occur [100], and Jackson and colleagues have suggested that PgsA2 might be the enzyme responsible for this reaction [9].

3.3.3. PE

The precise structure of PE was recently reported as 1-*O*-tuberculostearoyl-2-*O*-palmitoyl-*sn*-glycero-3-phosphoethanolamine in *M. tuberculosis* using tandem mass spectrometry [101]. In the initial step of PE synthesis, PS synthetase transfers serine to CDP-DAG and produces PS. In the second step, PE is produced by decarboxylation of PS mediated by PS decarboxylase. Genes encoding putative PS synthetase (*pssA*, *Rv0436c*) and PS decarboxylase (*psd*, *Rv0437c*) are found in tandem in the *M. tuberculosis* genome [9]. However, there is no experimental evidence demonstrating the identities of the genes. In *M. smegmatis*, PS synthetase and PS decarboxylase activities are enriched in different membrane fractions, which can be distinguished by sucrose gradient sedimentation [102]. However, the significance of differential membrane localization remains to be clarified.

3.3.4. PI

PI is a major phospholipid in both mycobacteria and corynebacteria and forms the anchor for the PIMs, which are substrates for heavy mannosylation to form LMs and additional arabinosylation to produce LAMs. PI is formed by the PI synthase PgsA (Rv2612c) from CDP-DAG and *myo*-inositol [9, 103]. Inositol is not a common metabolite in bacteria. It is therefore surprising that mycobacteria produce copious amounts of inositol through pathways shared with eukaryotes for incorporation into a range of metabolic pathways (recently reviewed in [104]). The enzyme D-*myo*-inositol 3-phosphate synthase (Ino1) converts glucose-6-phosphate to D-*myo*-inositol 3-phosphate, which is dephosphorylated by one of several inositol monophosphatases to produce *myo*-inositol [105].

3.3.5. PIMs

All Corynebacterineae synthesize PIMs that are important components of the cell envelope. Polar PIM species can also serve as membrane anchors for LM and LAM. Many of the steps of PIM/LM/LAM biosynthesis have now been elucidated [106]. Extensive genetic and biochemical studies have demonstrated that the synthesis of PIMs occurs linearly in mycobacteria with PI as the starting substrate (reviewed in [107]) (Fig. 4A). Early steps of the pathway occur on the cytoplasmic face of the plasma membrane. A PIM biosynthetic membrane, enriched in the early steps, has been purified by sucrose gradient fractionation as a membrane subdomain termed PMf, which is distinct from the bulk plasma membrane [102]. Mannosyltransferases performing the early steps utilize the water-soluble mannose donor, GDP-Man, which can be produced from exogenously acquired mannose or via *de novo* synthesis from the glycolytic pathway when fructose-6-phosphate is transformed by the enzymes ManA (Rv3255c), ManB (Rv3257c) and ManC (Rv3264c) [108-111], with a degree of redundancy reported at the ManC (NCgl0710) step in *C. glutamicum* [112]. The first step of PIM synthesis involves mannosylation of the C2-position of the inositol ring of PI by the enzyme PimA (Rv2610c) to form PIM1 [113, 114]. PimA is an essential enzyme in mycobacteria [11, 113] and absent in humans, making it a current target for drug development by several groups. The crystal structure of PimA from *M. smegmatis* has been solved in complex with GDP and GDP-Man at resolutions of 2.4Å and 2.6Å, respectively [115, 116]. PimA has a typical GT-B fold of glycosyltransferases consisting of two Rossmann-fold domains with a deep fissure at the interface, which contains the active site. Close to the GDP-Man binding site, the N-terminal domain displays a deep pocket containing highly conserved hydrophobic residues. This pocket is proposed to bind the acyl moieties of the acceptor substrate PI [116].

Next, O-6 mannosylation of the *myo*-inositol ring is performed by the cytoplasmic α-mannosyltransferase PimB' (Rv2188c) [117, 118] resulting in the formation of PIM2. While the mycobacterial enzyme Rv0557 was originally assigned this function [119], later studies showed that Rv2188c was the true PimB (designated PimB') [118], with Rv0557 being renamed MgtA and included in the LM-B pathway [120, 121] (see below). PimB' is an essential enzyme in mycobacteria [11, 117] and the crystal structure of the equivalent corynebacterial enzyme has been solved at high resolution complexed with nucleotide [122]. In corynebacteria, PIM2 is detected mainly in its mono-acylated form, but in mycobacteria PIM2 accumulates in the cell envelope as monoacyl (AcPIM2) and diacyl (Ac₂PIM2) forms, the former produced by the acyltransferase Rv2611c [123] which acts optimally on the product of the PimB' reaction [117].

The next enzyme in the pathway, PimC, has been identified in *M. tuberculosis* strain CDC1551 and could produce trimannosylated PIMs [124]. However, the absence of this enzyme in other strains indicates redundancy at this step and the enzymes involved remain to be identified, as does the putative "PimD" protein. Flipping of PIM intermediates from the cytoplasmic face of the membrane to the periplasm is thought to occur at this point of the pathway, but the precise intermediate and transporter involved are also undefined.

AcPIM4 species can be further mannosylated to form more polar PIMs in reactions thought to take place on the periplasmic side of the cytoplasmic membrane. These reactions are performed by glycosyltransferases that require a lipid sugar donor in the form of PPM, since these reactions are amphomycin-sensitive [36, 125-128]. In mycobacteria, AcPIM4 is proposed to be a branch point for synthesis of polar PIM end products and LM/LAM. PimE (Rv1159) has been shown to elongate AcPIM4 with one or more α1-2 linked mannoses to form AcPIM6 [36]. This polytopic membrane protein has sequence similarities with eukaryotic PIG-M mannosyltransferases and localizes to a cell wall-associated plasma membrane subdomain in *M. smegmatis*, termed PM-CW, to which enzymatic activities of AcPIM4-6 synthesis are enriched [102]. Its catalytic activity was successfully mapped to a conserved aspartate residue in the first outer loop of the protein. Whether PimE also forms AcPIM6 is unknown. Interestingly, no PimE orthologue is present in *C. glutamicum* and

Ac/Ac2PIM6 do not accumulate in this species, so the formation of Ac/Ac2PIM5/6 appears to be a mycobacterium-specific side-branch of the pathway.

Surprisingly, studies in *M. smegmatis* have revealed a role for a lipoprotein in PIM/LAM synthesis, since *lpqW* mutants produce reduced levels of LM/LAM [129]. As the only non-enzymatic component of the pathway identified to date, LpqW has been proposed to have a regulatory role at the bifurcation point of the pathways, as well as a functional connection with PimE, since mutations in the *pimE* gene can bypass the requirement for LpqW [130]. Very recent studies have implicated the corynebacterial ortholog of LpqW in the LM-B pathway as well (Rainczuk *et al*, submitted). Structural studies on the *M. smegmatis* LpqW revealed a scaffold similar to substrate binding proteins associated with ABC transporters, which has evolved to fulfill a new role in the regulation of PIM/LAM biosynthesis [131].

3.3.6. LM/LAM

A subpopulation of PIMs (AcPIM4 in mycobacteria [128, 129, 132] and AcPIM2 in corynebacteria [133]) can be extended with chains of α1-6 linked mannose to form LM that is further modified with a number of single α1-2 mannose side chains [134-136]. MptB is a PPM-dependent mannosyltransferase involved in extending AcPIM2 to form the proximal α1-6 mannan backbone of LM in *C. glutamicum* but is redundant in *M. smegmatis* [133], indicating additional complexity at this step in mycobacteria. Further elongation is performed by MptA (Rv2174) [135, 136], with MptC (Rv2181) required for addition of α1-2 linked Man side chains [106, 134, 137, 138]. Further additions of arabinose units by EmbC (Rv3793), AftC (Rv2673), AftD (Rv0236c) and at present unidentified α1-5 arabinofuranosyltransferases, result in the formation of mature LAM [139-142]. In *M. tuberculosis* and other pathogenic mycobacteria, additional mannose capping is present [143], synthesized by the enzymes MptC [137] and Rv1635c [144]. Alternatively, *M. smegmatis* LAM is capped with inositol phosphate [145].

While the general PI→PIM→LM→LAM pathway is conserved in corynebacteria, a second pathway of lipoglycan biosynthesis exists in which a sub-population of LM lipoglycans is assembled on a glucopyranosyluronic acid diacylglycerol (Gl-A, GlcADAG) glycolipid anchor [118, 121, 133]. In this pathway (Fig. 4B), Gl-A is first mannosylated by MgtA (NCgl0452 in *C. glutamicum* and previously termed PimB, see above) forming mannosyl-glucuronic acid diacylglycerol (Gl-X, ManGlcADAG), which is subjected to further α1-6 (backbone) and α1-2 (side unit) mannosylation resulting in LM [121]. This pathway shares some PPM-dependent enzymes with the PI-based LM pathway including MptB, since an *mptB* (*NCgl1505*) mutant of *C. glutamicum* fails to produce intermediates beyond Gl-X or AcPIM2 [133]. For clarity, the PI-based LM pool has been designated LM-A while this second pathway produces LM-B [121, 146], the major LM pool in *C. glutamicum* [118]. While *C. glutamicum* produces LAMs via LM-A using a similar pathway to mycobacteria, its LAMs are smaller and structurally distinct, with more extensive mannosylation and singular Ara*f* capping [147].

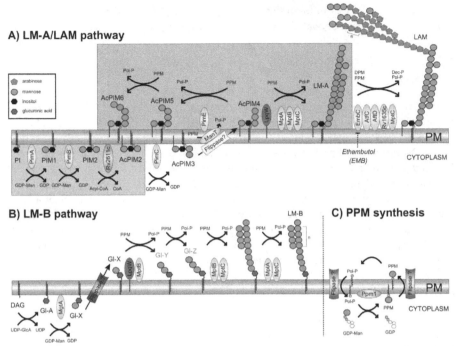

Figure 4. Summary of Lipoglycan Biosynthesis Pathways in Corynebacterineae. A) The mycobacterial pathway and structures for LM-A/LAM are shown, although several steps are inferred from studies in *C. glutamicum*. A progression from PIM1 to AcPIM2 via AcPIM1 can also occur but is sub-optimal in mycobacteria. AcPIM3 is shown as the substrate for the unidentified flippase but AcPIM2 may be the preferred substrate: AcPIM2 appears to be the flipped intermediate in *C. glutamicum*. Reactions associated with the PMf and PM-CW subfractions of *M. smegmatis* are shaded orange and grey, respectively. B) LM-B pathway in *C. glutamicum*. Its presence in mycobacteria remains to be determined. Intermediates representing Gl-Y and Gl-Z have been detected in very recent studies (Rainczuk et al, submitted). C) PPM serves as the lipid-linked donor of mannose for both pathways. The mechanism of flipping is undetermined. PM, plasma membrane. Other abbreviations are as defined in the text.

3.4. Prenol lipids

Polyprenol phosphate (Pol-P) is a key carrier lipid in synthesis of the core structures of the mycobacterial cell wall, including peptidoglycan and arabinogalactan. Unlike most bacteria, mycobacteria contain multiple types of Pol-P. For example, *M. smegmatis* produces decaprenol phosphate (C_{50}, Dec-P, [148]) and heptaprenol phosphate (C_{35}, Hep-P, [149]). Polyprenols are thought to be synthesized via the condensation of two C_5 lipids, isopentenyl diphosphate [150-152] and dimethylallyl diphosphate derived from the mevalonate-independent methylerythritol 4-phosphate (MEP) pathway. These reactions are catalyzed by prenol diphosphate synthases and two *M. tuberculosis* proteins, Rv2361c and Rv1086, acting sequentially, are thought to fulfill this role [148, 149, 153-155]. Finally, the C_{50} decaprenol diphosphate is dephosphorylated to produce Dec-P by an unknown phosphatase.

3.4.1. Polyprenol-phospho-sugars

3.4.1.1. PPM

PPM, a β-D-mannosyl-1-monophosphoryldecaprenol, is utilized by periplasmic mannosyltransferases for synthesis of polar PIM species and LMs [106]. C_{35}/C_{50}-P-Man$p$ is formed by the PPM synthase Ppm1 (Rv2051c) from GDP-Man$p$ and Pol-P [126, 127] (Fig. 4C). In *M. tuberculosis* Ppm1 consists of two domains: a C-terminal catalytic domain and a N-terminal membrane anchor with 6 transmembrane helices. However, the equivalent domains of the *M. smegmatis* PPM synthase are two distinct, but interacting, proteins [156]. The importance of PPM in cell wall synthesis has been highlighted by analysis of a *ppm1* mutant in *C. glutamicum* that failed to produce PPM, resulting in severe defects in lipoglycan biosynthesis. While the mutant could synthesize PIM2 species, all downstream products (LM, LAM) were absent, indicating their reliance on the PPM donor [127]. These findings were consistent with others showing that amphomycin, an antibiotic specific for PPM-dependent polymerases, blocked the PIM pathway at the PIM2 or PIM3 stage [125, 128]. The product of the *Rv3779* gene has also been implicated in PPM synthesis but its role remains unclear [157, 158].

3.4.1.2. DPA

DPA is the only known donor of arabinose (Ara) for mycobacterial cell wall synthesis, contributing Ara$f$ units to arabinogalactan and LAM [106] with concomitant release of Pol-P. The Ara$f$ portion is derived from the pentose-phosphate pathway [159-161]. 5-phosphoribose 1-diphosphate is transferred to Dec-P by the *Rv3806c* gene product [162] and the resultant Dec-P-β-D-5-phosphoribose is dephosphorylated to form Dec-P-β-D-ribose. Oxidation of the 2′ hydroxyl is followed by a reduction reaction to form DPA. This two-step epimerization reaction is catalyzed by the combined activities of DprE1 (Rv3790) and DprE2 (Rv3791) [163]. Since DPA is the sole donor of Ara$f$ residues for mycobacterial cell wall synthesis, this pathway is of interest for drug development. Indeed, DprE1 is the target of dinitrobenzamide derivatives (DNBs) [164] and a set of nitro-compounds related to DNBs, the nitro-benzothiazinones (BTZ), a class of compounds with nanomolar anti-*M. tuberculosis* activities but minimal host-cell toxicity [165-167]. The essential nature of DprE1 in species beyond *M. tuberculosis* [168] reinforces it as a "magic" drug target [169]. The crystal structure of *M. tuberculosis* DprE1 complexed with BTZ inhibitors has been reported very recently, revealing the mode of inhibitor binding [170].

3.4.2. Carotenoids

Carotenoids are isoprenoid pigments widely distributed in biology and mostly based on C_{40}-polyene. Synthesis of these pigments has been poorly studied in mycobacteria but they have proven useful for taxonomic and identification purposes. Mycobacterial pigments are generally yellow or orange and most have been confirmed as carotenoids [171]. While carotenoid genetics has been best studied in plants, the key enzymes of the pathway have been identified in bacteria, including mycobacteria [172-174]. There are two classes of carotenoids in bacteria, carotenes and xanthophylls, the latter of which contain oxygen. Both classes are composed of eight isoprenoid units with a long central chain of double bonded carbons. A consensus pathway for carotenoid biosynthesis in bacteria has been elucidated with orthologues of key enzymes identified in several species of mycobacteria [175]. As described above for prenol lipids, the carotenoid pathway begins with isopentenyl diphosphate and dimethylallyl diphosphate derived from the MEP pathway [176]. Head-to-tail condensation of these terpenes produces geranylgeranyl pyrophosphate (GGPP) due to the activity of GGPP synthase (CrtE). Condensation of two GGPP molecules [177] is followed by desaturation to phytoene by phytoene synthase (CrtB). Phytoene desaturase (CrtI) converts phytoene to lycoprene followed by cyclization to β-carotene by lycoprene cyclase (CrtY) [178].

4. Concluding remarks

Lipid metabolism in mycobacteria and corynebacteria is a highly complex network of catabolic and anabolic reactions. While the metabolic pathways and many of the enzymes involved have been actively elucidated over the past decade, substantial efforts are still needed to draw a comprehensive map of lipid metabolism in these organisms. In particular, our understanding of regulatory mechanisms of lipid metabolism is currently at an early stage. In addition, there are very few studies describing the interactions between multiple metabolic pathways of lipid biosynthesis. One promising approach for the comprehensive understanding of lipid metabolism is lipidomics, which is the study of lipid biosynthetic and catabolic pathways at a global level [179]. In the past, metabolic pathways have generally been examined in isolation without consideration of how different pathways might interact with, and influence, one another. Since the plasma membrane is a shared platform for most lipid biosynthetic pathways, and some donors are shared between different pathways (see above), it seems unlikely that the various pathways are truly independent. Recent advances in mass spectrometry (*e.g.* MALDI-MS, ESI-MS), nuclear magnetic resonance spectroscopy and associated computational methods have fuelled the development of this field [180]. Members of the Corynebacterineae, with their extensive lipid repertoires and complex metabolic pathways, would seem to be ideal targets to assess the true potential of lipidomics technologies. Recently, appropriate databases and methods for detection and identification of all major lipid classes of *M. tuberculosis* from a single crude extract have been developed [181, 182]. Very recently, a lipidomics profiling platform has been reported that uses high-performance liquid chromatography/mass spectrometry to resolve more than 12,000 molecules from *M. tuberculosis* [183]. These exciting advances provide a basis for future studies on the regulation of lipid metabolism and may allow, for the first time, a true appreciation of the interactive lipid networks of mycobacteria and corynebacteria.

Author details

Paul K. Crellin

Australian Research Council Centre of Excellence in Structural and Functional Microbial Genomics, Department of Microbiology, Monash University, Australia

Chu-Yuan Luo and Yasu S. Morita*

Department of Microbiology, University of Massachusetts, Amherst, USA

5. References

[1] 1.   Hoffmann C, Leis A, Niederweis M, Plitzko JM, Engelhardt H. Disclosure of the mycobacterial outer membrane: cryo-electron tomography and vitreous sections reveal the lipid bilayer structure. Proc Natl Acad Sci U S A. 2008;105:3963-7.

[2] Zuber B, Chami M, Houssin C, Dubochet J, Griffiths G, Daffe M. Direct visualization of the outer membrane of mycobacteria and corynebacteria in their native state. J Bacteriol. 2008;190:5672-80.

[3] Barry CE, 3rd, Mdluli K. Drug sensitivity and environmental adaptation of mycobacterial cell wall components. Trends Microbiol. 1996;4:275-81.

[4] Daffe M, Draper P. The envelope layers of mycobacteria with reference to their pathogenicity. Adv Microb Physiol. 1998;39:131-203.

[5] Cole ST, Brosch R, Parkhill J, Garnier T, Churcher C, Harris D, et al. Deciphering the biology of *Mycobacterium tuberculosis* from the complete genome sequence. Nature. 1998;393:537-44.

[6] Nandedkar AK. Comparative study of the lipid composition of particular pathogenic and nonpathogenic species of *Mycobacterium*. J Natl Med Assoc. 1983;75:69-74.

[7] Subramoniam A, Subrahmanyam D. Light-induced changes in the phospholipid composition of *Mycobacterium smegmatis* ATCC 607. J Gen Microbiol. 1982;128:419-21.

[8] Khuller GK, Taneja R, Nath N. Effect of fatty acid supplementation on the lipid composition of *Mycobacterium smegmatis* ATCC 607, grown at 27 degrees and 37 degrees C. J Appl Bacteriol. 1983;54:63-8.

[9] Jackson M, Crick DC, Brennan PJ. Phosphatidylinositol is an essential phospholipid of mycobacteria. J Biol Chem. 2000;275:30092-9.

[10] Akamatsu Y, Nojima S. Separation and analyses of the individual phospholipids of mycobacteria. J Biochem. 1965;57:430-9.

[11] Sassetti CM, Boyd DH, Rubin EJ. Genes required for mycobacterial growth defined by high density mutagenesis. Mol Microbiol. 2003;48:77-84.

[12] Lamichhane G, Zignol M, Blades NJ, Geiman DE, Dougherty A, Grosset J, et al. A postgenomic method for predicting essential genes at subsaturation levels of mutagenesis: application to *Mycobacterium tuberculosis*. Proc Natl Acad Sci U S A. 2003;100:7213-8.

---

* Corresponding Author

[13] Yague G, Segovia M, Valero-Guillen PL. Phospholipid composition of several clinically relevant *Corynebacterium* species as determined by mass spectrometry: an unusual fatty acyl moiety is present in inositol-containing phospholipids of *Corynebacterium urealyticum*. Microbiology. 2003;149:1675-85.

[14] Mileykovskaya E, Dowhan W, Birke RL, Zheng D, Lutterodt L, Haines TH. Cardiolipin binds nonyl acridine orange by aggregating the dye at exposed hydrophobic domains on bilayer surfaces. FEBS Lett. 2001;507:187-90.

[15] Maloney E, Madiraju SC, Rajagopalan M, Madiraju M. Localization of acidic phospholipid cardiolipin and DnaA in mycobacteria. Tuberculosis (Edinb). 2011;91 Suppl 1:S150-5.

[16] Maloney E, Lun S, Stankowska D, Guo H, Rajagoapalan M, Bishai WR, et al. Alterations in phospholipid catabolism in *Mycobacterium tuberculosis* lysX mutant. Front Microbiol. 2011;2:19.

[17] Dahlberg M. Polymorphic phase behavior of cardiolipin derivatives studied by coarse-grained molecular dynamics. J Phys Chem B. 2007;111:7194-200.

[18] Powell GL, Marsh D. Polymorphic phase behavior of cardiolipin derivatives studied by 31P NMR and X-ray diffraction. Biochemistry. 1985;24:2902-8.

[19] Lewis RN, Zweytick D, Pabst G, Lohner K, McElhaney RN. Calorimetric, x-ray diffraction, and spectroscopic studies of the thermotropic phase behavior and organization of tetramyristoyl cardiolipin membranes. Biophys J. 2007;92:3166-77.

[20] Imai T, Kageyama Y, Tobari J. *Mycobacterium smegmatis* malate dehydrogenase: activation of the lipid-depleted enzyme by anionic phospholipids and phosphatidylethanolamine. Biochim Biophys Acta. 1995;1246:189-96.

[21] Kimura T, Tobari J. Participation of flavin-adenine dinucleotide in the activity of malate dehydrogenase from *Mycobacterium avium*. Biochim Biophys Acta. 1963;73:399-405.

[22] Tobari J. Requirement of flavin adenine dinucleotide and phospholipid for the activity of malate dehydrogenase from *Mycobacterium avium*. Biochem Biophys Res Commun. 1964;15:50-4.

[23] Marchand CH, Salmeron C, Bou Raad R, Meniche X, Chami M, Masi M, et al. Biochemical disclosure of the mycolate outer membrane of *Corynebacterium glutamicum*. J Bacteriol. 2012;194:587-97.

[24] Bansal-Mutalik R, Nikaido H. Quantitative lipid composition of cell envelopes of *Corynebacterium glutamicum* elucidated through reverse micelle extraction. Proc Natl Acad Sci U S A. 2011;108:15360-5.

[25] Fischer K, Chatterjee D, Torrelles J, Brennan PJ, Kaufmann SH, Schaible UE. Mycobacterial lysocardiolipin is exported from phagosomes upon cleavage of cardiolipin by a macrophage-derived lysosomal phospholipase A2. J Immunol. 2001;167:2187-92.

[26] Murzyn K, Rog T, Pasenkiewicz-Gierula M. Phosphatidylethanolamine-phosphatidylglycerol bilayer as a model of the inner bacterial membrane. Biophys J. 2005;88:1091-103.

[27] Brennan PJ, Lehane DP. The phospholipids of corynebacteria. Lipids. 1971;6:401-9.

[28] Khuller GK, Brennan PJ. Further studies on the lipids of corynebacteria. The mannolipids of *Corynebacterium aquaticum*. Biochem J. 1972;127:369-73.

[29] Suzuki KI, Suzuki M, Sasaki J, Park YH, Komagata KK. Leifsonia gen. nov., a genus for 2,4-diaminobutyric acid-containing actinomycetes to accommodate "*Corynebacterium aquaticum*" Leifson 1962 and Clavibacter xyli subsp. cynodontis Davis et al. 1984. J Gen Appl Microbiol. 1999;45:253-62.

[30] Charalambous K, Miller D, Curnow P, Booth PJ. Lipid bilayer composition influences small multidrug transporters. BMC Biochem. 2008;9:31.

[31] Chatterjee D, Khoo KH. Mycobacterial lipoarabinomannan: an extraordinary lipoheteroglycan with profound physiological effects. Glycobiology. 1998;8:113-20.

[32] Nigou J, Gilleron M, Rojas M, Garcia LF, Thurnher M, Puzo G. Mycobacterial lipoarabinomannans: modulators of dendritic cell function and the apoptotic response. Microbes Infect. 2002;4:945-53.

[33] Strohmeier GR, Fenton MJ. Roles of lipoarabinomannan in the pathogenesis of tuberculosis. Microbes Infect. 1999;1:709-17.

[34] Vercellone A, Nigou J, Puzo G. Relationships between the structure and the roles of lipoarabinomannans and related glycoconjugates in tuberculosis pathogenesis. Front Biosci. 1998;3:e149-63.

[35] Pitarque S, Larrouy-Maumus G, Payre B, Jackson M, Puzo G, Nigou J. The immunomodulatory lipoglycans, lipoarabinomannan and lipomannan, are exposed at the mycobacterial cell surface. Tuberculosis (Edinb). 2008;88:560-5.

[36] Morita YS, Sena CB, Waller RF, Kurokawa K, Sernee MF, Nakatani F, et al. PimE is a polyprenol-phosphate-mannose-dependent mannosyltransferase that transfers the fifth mannose of phosphatidylinositol mannoside in mycobacteria. J Biol Chem. 2006;281:25143-55.

[37] Lechevalier MP, Debievre C, Lechevalier H. Chemotaxonomy of aerobic actinomycetes - phospholipid composition. Biochem Syst Ecol. 1977;5:249-60.

[38] Yague G, Segovia M, Valero-Guillen PL. Acyl phosphatidylglycerol: a major phospholipid of *Corynebacterium amycolatum*. FEMS Microbiol Lett. 1997;151:125-30.

[39] Niepel T, Meyer H, Wray V, Abraham WR. Intraspecific variation of unusual phospholipids from *Corynebacterium* spp. containing a novel fatty acid. J Bacteriol. 1998;180:4650-7.

[40] Mazzella N, Molinet J, Syakti AD, Dodi A, Doumenq P, Artaud J, et al. Bacterial phospholipid molecular species analysis by ion-pair reversed-phase HPLC/ESI/MS. J Lipid Res. 2004;45:1355-63.

[41] Valero-Guillen PL, Yague G, Segovia M. Characterization of acyl-phosphatidylinositol from the opportunistic pathogen *Corynebacterium amycolatum*. Chem Phys Lipids. 2005;133:17-26.

[42] Puech V, Chami M, Lemassu A, Laneelle MA, Schiffler B, Gounon P, et al. Structure of the cell envelope of corynebacteria: importance of the non-covalently bound lipids in the formation of the cell wall permeability barrier and fracture plane. Microbiology. 2001;147:1365-82.

[43] Morita YS, Yamaryo-Botte Y, Miyanagi K, Callaghan JM, Patterson JH, Crellin PK, et al. Stress-induced synthesis of phosphatidylinositol 3-phosphate in mycobacteria. J Biol Chem. 2010;285:16643-50.

[44] Maloney E, Stankowska D, Zhang J, Fol M, Cheng QJ, Lun S, et al. The two-domain LysX protein of *Mycobacterium tuberculosis* is required for production of lysinylated phosphatidylglycerol and resistance to cationic antimicrobial peptides. PLoS Pathog. 2009;5:e1000534.

[45] Fraser PD, Bramley PM. The biosynthesis and nutritional uses of carotenoids. Prog Lipid Res. 2004;43:228-65.

[46] Scherzinger D, Scheffer E, Bar C, Ernst H, Al-Babili S. The *Mycobacterium tuberculosis* ORF *Rv0654* encodes a carotenoid oxygenase mediating central and excentric cleavage of conventional and aromatic carotenoids. FEBS J. 2010;277:4662-73.

[47] Steinbuchel A, Valentin HE. Diversity of bacterial polyhydroxyalkanoic acids. FEMS Microbiol Lett. 1995;128:219-28.

[48] Daniel J, Deb C, Dubey VS, Sirakova TD, Abomoelak B, Morbidoni HR, et al. Induction of a novel class of diacylglycerol acyltransferases and triacylglycerol accumulation in *Mycobacterium tuberculosis* as it goes into a dormancy-like state in culture. J Bacteriol. 2004;186:5017-30.

[49] Alvarez HM, Steinbuchel A. Triacylglycerols in prokaryotic microorganisms. Appl Microbiol Biotechnol. 2002;60:367-76.

[50] Alvarez HM, Kalscheuer R, Steinbuchel A. Accumulation and mobilization of storage lipids by *Rhodococcus opacus* PD630 and *Rhodococcus ruber* NCIMB 40126. Appl Microbiol Biotechnol. 2000;54:218-23.

[51] Alvarez HM, Luftmann H, Silva RA, Cesari AC, Viale A, Waltermann M, et al. Identification of phenyldecanoic acid as a constituent of triacylglycerols and wax ester produced by *Rhodococcus opacus* PD630. Microbiology. 2002;148:1407-12.

[52] Olukoshi ER, Packter NM. Importance of stored triacylglycerols in *Streptomyces*: possible carbon source for antibiotics. Microbiology. 1994;140 (Pt 4):931-43.

[53] Daniel J, Maamar H, Deb C, Sirakova TD, Kolattukudy PE. *Mycobacterium tuberculosis* uses host triacylglycerol to accumulate lipid droplets and acquires a dormancy-like phenotype in lipid-loaded macrophages. PLoS Pathog. 2011;7:e1002093.

[54] Sirakova TD, Dubey VS, Deb C, Daniel J, Korotkova TA, Abomoelak B, et al. Identification of a diacylglycerol acyltransferase gene involved in accumulation of triacylglycerol in *Mycobacterium tuberculosis* under stress. Microbiology. 2006;152:2717-25.

[55] Pandey AK, Sassetti CM. Mycobacterial persistence requires the utilization of host cholesterol. Proc Natl Acad Sci U S A. 2008;105:4376-80.

[56] Griffin JE, Gawronski JD, Dejesus MA, Ioerger TR, Akerley BJ, Sassetti CM. High-resolution phenotypic profiling defines genes essential for mycobacterial growth and cholesterol catabolism. PLoS Pathog. 2011;7:e1002251.

[57] Brzostek A, Pawelczyk J, Rumijowska-Galewicz A, Dziadek B, Dziadek J. *Mycobacterium tuberculosis* is able to accumulate and utilize cholesterol. J Bacteriol. 2009;191:6584-91.

[58] Av-Gay Y, Sobouti R. Cholesterol is accumulated by mycobacteria but its degradation is limited to non-pathogenic fast-growing mycobacteria. Can J Microbiol. 2000;46:826-31.

[59] Wright LD, Skeggs HR. Reversal of sodium propionate inhibition of *Escherichia coli* with beta-alanine. Arch Biochem. 1946;10:383-6.

[60] Brock M, Buckel W. On the mechanism of action of the antifungal agent propionate. Eur J Biochem. 2004;271:3227-41.

[61] Maruyama K, Kitamura H. Mechanisms of growth inhibition by propionate and restoration of the growth by sodium bicarbonate or acetate in *Rhodopseudomonas sphaeroides* S. J Biochem. 1985;98:819-24.

[62] Savvi S, Warner DF, Kana BD, McKinney JD, Mizrahi V, Dawes SS. Functional characterization of a vitamin B12-dependent methylmalonyl pathway in *Mycobacterium tuberculosis*: implications for propionate metabolism during growth on fatty acids. J Bacteriol. 2008;190:3886-95.

[63] Singh A, Crossman DK, Mai D, Guidry L, Voskuil MI, Renfrow MB, et al. *Mycobacterium tuberculosis* WhiB3 maintains redox homeostasis by regulating virulence lipid anabolism to modulate macrophage response. PLoS Pathog. 2009;5:e1000545.

[64] Bloch K. Control mechanisms for fatty acid synthesis in *Mycobacterium smegmatis*. Adv Enzymol Relat Areas Mol Biol. 1977;45:1-84.

[65] Fernandes ND, Kolattukudy PE. Cloning, sequencing and characterization of a fatty acid synthase-encoding gene from *Mycobacterium tuberculosis* var. bovis BCG. Gene. 1996;170:95-9.

[66] Bloch K, Vance D. Control mechanisms in the synthesis of saturated fatty acids. Annu Rev Biochem. 1977;46:263-98.

[67] Radmacher E, Alderwick LJ, Besra GS, Brown AK, Gibson KJ, Sahm H, et al. Two functional FAS-I type fatty acid synthases in *Corynebacterium glutamicum*. Microbiology. 2005;151:2421-7.

[68] Odriozola JM, Ramos JA, Bloch K. Fatty acid synthetase activity in *Mycobacterium smegmatis*. Characterization of the acyl carrier protein-dependent elongating system. Biochim Biophys Acta. 1977;488:207-17.

[69] Rock CO, Cronan JE. *Escherichia coli* as a model for the regulation of dissociable (type II) fatty acid biosynthesis. Biochim Biophys Acta. 1996;1302:1-16.

[70] Kremer L, Nampoothiri KM, Lesjean S, Dover LG, Graham S, Betts J, et al. Biochemical characterization of acyl carrier protein (AcpM) and malonyl-CoA:AcpM transacylase (mtFabD), two major components of *Mycobacterium tuberculosis* fatty acid synthase II. J Biol Chem. 2001;276:27967-74.

[71] Marrakchi H, Laneelle G, Quemard A. InhA, a target of the antituberculous drug isoniazid, is involved in a mycobacterial fatty acid elongation system, FAS-II. Microbiology. 2000;146:289-96.

[72] Portevin D, De Sousa-D'Auria C, Houssin C, Grimaldi C, Chami M, Daffe M, et al. A polyketide synthase catalyzes the last condensation step of mycolic acid biosynthesis in mycobacteria and related organisms. Proc Natl Acad Sci U S A. 2004;101:314-9.

[73] Lea-Smith DJ, Pyke JS, Tull D, McConville MJ, Coppel RL, Crellin PK. The reductase that catalyzes mycolic motif synthesis is required for efficient attachment of mycolic acids to arabinogalactan. J Biol Chem. 2007;282:11000-8.

[74] Athenstaedt K, Daum G. Phosphatidic acid, a key intermediate in lipid metabolism. Eur J Biochem. 1999;266:1-16.

[75] Okuyama H, Kameyama Y, Fujikawa M, Yamada K, Ikezawa H. Mechanism of diacylglycerophosphate synthesis in mycobacteria. J Biol Chem. 1977;252:6682-6.

[76] Okuyama H, Kankura T, Nojima S. Positional distribution of fatty acids in phospholipids from mycobacteria. J Biochem. 1967;61:732-7.

[77] Owens RM, Hsu FF, VanderVen BC, Purdy GE, Hesteande E, Giannakas P, et al. M. tuberculosis Rv2252 encodes a diacylglycerol kinase involved in the biosynthesis of phosphatidylinositol mannosides (PIMs). Mol Microbiol. 2006;60:1152-63.

[78] Smith SW, Weiss SB, Kennedy EP. The enzymatic dephosphorylation of phosphatidic acids. J Biol Chem. 1957;228:915-22.

[79] Han GS, Wu WI, Carman GM. The Saccharomyces cerevisiae Lipin homolog is a Mg^{2+}-dependent phosphatidate phosphatase enzyme. J Biol Chem. 2006;281:9210-8.

[80] Kalscheuer R, Steinbuchel A. A novel bifunctional wax ester synthase/acyl-CoA:diacylglycerol acyltransferase mediates wax ester and triacylglycerol biosynthesis in Acinetobacter calcoaceticus ADP1. J Biol Chem. 2003;278:8075-82.

[81] Elamin AA, Stehr M, Spallek R, Rohde M, Singh M. The Mycobacterium tuberculosis Ag85A is a novel diacylglycerol acyltransferase involved in lipid body formation. Mol Microbiol. 2011;81:1577-92.

[82] Ortalo-Magne A, Lemassu A, Laneelle MA, Bardou F, Silve G, Gounon P, et al. Identification of the surface-exposed lipids on the cell envelopes of Mycobacterium tuberculosis and other mycobacterial species. J Bacteriol. 1996;178:456-61.

[83] Paznokas JL, Kaplan A. Purification and properties of a triacylglycerol lipase from Mycobacterium phlei. Biochim Biophys Acta. 1977;487:405-21.

[84] Deb C, Daniel J, Sirakova TD, Abomoelak B, Dubey VS, Kolattukudy PE. A novel lipase belonging to the hormone-sensitive lipase family induced under starvation to utilize stored triacylglycerol in Mycobacterium tuberculosis. J Biol Chem. 2006;281:3866-75.

[85] Mishra KC, de Chastellier C, Narayana Y, Bifani P, Brown AK, Besra GS, et al. Functional role of the PE domain and immunogenicity of the Mycobacterium tuberculosis triacylglycerol hydrolase LipY. Infect Immun. 2008;76:127-40.

[86] Munoz-Elias EJ, McKinney JD. Carbon metabolism of intracellular bacteria. Cell Microbiol. 2006;8:10-22.

[87] Cotes K, Dhouib R, Douchet I, Chahinian H, de Caro A, Carriere F, et al. Characterization of an exported monoglyceride lipase from Mycobacterium tuberculosis possibly involved in the metabolism of host cell membrane lipids. Biochem J. 2007;408:417-27.

[88] Waltermann M, Steinbuchel A. Neutral lipid bodies in prokaryotes: recent insights into structure, formation, and relationship to eukaryotic lipid depots. J Bacteriol. 2005;187:3607-19.

[89] Waltermann M, Hinz A, Robenek H, Troyer D, Reichelt R, Malkus U, et al. Mechanism of lipid-body formation in prokaryotes: how bacteria fatten up. Mol Microbiol. 2005;55:750-63.

[90] Hanisch J, Waltermann M, Robenek H, Steinbuchel A. Eukaryotic lipid body proteins in oleogenous actinomycetes and their targeting to intracellular triacylglycerol inclusions: Impact on models of lipid body biogenesis. Appl Environ Microbiol. 2006;72:6743-50.

[91] Hanisch J, Waltermann M, Robenek H, Steinbuchel A. The *Ralstonia eutropha* H16 phasin PhaP1 is targeted to intracellular triacylglycerol inclusions in *Rhodococcus opacus* PD630 and *Mycobacterium smegmatis* mc²155, and provides an anchor to target other proteins. Microbiology. 2006;152:3271-80.

[92] Dowhan W. CDP-diacylglycerol synthase of microorganisms. Biochim Biophys Acta. 1997;1348:157-65.

[93] Nigou J, Besra GS. Cytidine diphosphate-diacylglycerol synthesis in *Mycobacterium smegmatis*. Biochem J. 2002;367:157-62.

[94] Lecocq J, Ballou CE. On the structure of cardiolipin. Biochemistry. 1964;3:976-80.

[95] Icho T, Raetz CR. Multiple genes for membrane-bound phosphatases in *Escherichia coli* and their action on phospholipid precursors. J Bacteriol. 1983;153:722-30.

[96] Funk CR, Zimniak L, Dowhan W. The pgpA and pgpB genes of *Escherichia coli* are not essential: evidence for a third phosphatidylglycerophosphate phosphatase. J Bacteriol. 1992;174:205-13.

[97] Lu YH, Guan Z, Zhao J, Raetz CR. Three phosphatidylglycerol-phosphate phosphatases in the inner membrane of *Escherichia coli*. J Biol Chem. 2011;286:5506-18.

[98] Osman C, Haag M, Wieland FT, Brugger B, Langer T. A mitochondrial phosphatase required for cardiolipin biosynthesis: the PGP phosphatase Gep4. EMBO J. 2010;29:1976-87.

[99] Zhang J, Guan Z, Murphy AN, Wiley SE, Perkins GA, Worby CA, et al. Mitochondrial phosphatase PTPMT1 is essential for cardiolipin biosynthesis. Cell Metab. 2011;13:690-700.

[100] Mathur AK, Murthy PS, Saharia GS, Venkitasubramanian TA. Studies on cardiolipin biosynthesis in *Mycobacterium smegmatis*. Can J Microbiol. 1976;22:354-8.

[101] Ter Horst B, Seshadri C, Sweet L, Young DC, Feringa BL, Moody DB, et al. Asymmetric synthesis and structure elucidation of a glycerophospholipid from *Mycobacterium tuberculosis*. J Lipid Res. 2010;51:1017-22.

[102] Morita YS, Velasquez R, Taig E, Waller RF, Patterson JH, Tull D, et al. Compartmentalization of lipid biosynthesis in mycobacteria. J Biol Chem. 2005;280:21645-52.

[103] Salman M, Lonsdale JT, Besra GS, Brennan PJ. Phosphatidylinositol synthesis in mycobacteria. Biochim Biophys Acta. 1999;1436:437-50.

[104] Morita YS, Fukuda T, Sena CB, Yamaryo-Botte Y, McConville MJ, Kinoshita T. Inositol lipid metabolism in mycobacteria: biosynthesis and regulatory mechanisms. Biochim Biophys Acta. 2011;1810:630-41.

[105] Nigou J, Besra GS. Characterization and regulation of inositol monophosphatase activity in *Mycobacterium smegmatis*. Biochem J. 2002;361:385-90.

[106] Mishra AK, Driessen NN, Appelmelk BJ, Besra GS. Lipoarabinomannan and related glycoconjugates: structure, biogenesis and role in *Mycobacterium tuberculosis* physiology and host-pathogen interaction. FEMS Microbiol Rev. 2011;35:1126-57.

[107] Guerin ME, Kordulakova J, Alzari PM, Brennan PJ, Jackson M. Molecular basis of phosphatidyl-*myo*-inositol mannoside biosynthesis and regulation in mycobacteria. J Biol Chem. 2010;285:33577-83.

[108] Ma Y, Stern RJ, Scherman MS, Vissa VD, Yan W, Jones VC, et al. Drug targeting *Mycobacterium tuberculosis* cell wall synthesis: genetics of dTDP-rhamnose synthetic enzymes and development of a microtiter plate-based screen for inhibitors of conversion of dTDP-glucose to dTDP-rhamnose. Antimicrob Agents Chemother. 2001;45:1407-16.

[109] McCarthy TR, Torrelles JB, MacFarlane AS, Katawczik M, Kutzbach B, Desjardin LE, et al. Overexpression of *Mycobacterium tuberculosis* manB, a phosphomannomutase that increases phosphatidylinositol mannoside biosynthesis in *Mycobacterium smegmatis* and mycobacterial association with human macrophages. Mol Microbiol. 2005;58:774-90.

[110] Ning B, Elbein AD. Purification and properties of mycobacterial GDP-mannose pyrophosphorylase. Arch Biochem Biophys. 1999;362:339-45.

[111] Patterson JH, Waller RF, Jeevarajah D, Billman-Jacobe H, McConville MJ. Mannose metabolism is required for mycobacterial growth. Biochem J. 2003;372:77-86.

[112] Mishra AK, Krumbach K, Rittmann D, Batt SM, Lee OY, De S, et al. Deletion of manC in *Corynebacterium glutamicum* results in a phospho-myo-inositol mannoside- and lipoglycan-deficient mutant. Microbiology. 2012;158:1908-17.

[113] Kordulakova J, Gilleron M, Mikusova K, Puzo G, Brennan PJ, Gicquel B, et al. Definition of the first mannosylation step in phosphatidylinositol mannoside synthesis. PimA is essential for growth of mycobacteria. J Biol Chem. 2002;277:31335-44.

[114] Gu X, Chen M, Wang Q, Zhang M, Wang B, Wang H. Expression and purification of a functionally active recombinant GDP-mannosyltransferase (PimA) from *Mycobacterium tuberculosis* H37Rv. Protein Expr Purif. 2005;42:47-53.

[115] Guerin ME, Buschiazzo A, Kordulakova J, Jackson M, Alzari PM. Crystallization and preliminary crystallographic analysis of PimA, an essential mannosyltransferase from *Mycobacterium smegmatis*. Acta Crystallograph Sect F Struct Biol Cryst Commun. 2005;61:518-20.

[116] Guerin ME, Kordulakova J, Schaeffer F, Svetlikova Z, Buschiazzo A, Giganti D, et al. Molecular recognition and interfacial catalysis by the essential phosphatidylinositol mannosyltransferase PimA from mycobacteria. J Biol Chem. 2007;282:20705-14.

[117] Guerin ME, Kaur D, Somashekar BS, Gibbs S, Gest P, Chatterjee D, et al. New insights into the early steps of phosphatidylinositol mannoside biosynthesis in mycobacteria: PimB' is an essential enzyme of *Mycobacterium smegmatis*. J Biol Chem. 2009;284:25687-96.

[118] Lea-Smith DJ, Martin KL, Pyke JS, Tull D, McConville MJ, Coppel RL, et al. Analysis of a new mannosyltransferase required for the synthesis of phosphatidylinositol

mannosides and lipoarabinomannan reveals two lipomannan pools in *Corynebacterineae*. J Biol Chem. 2008;283:6773-82.

[119] Schaeffer ML, Khoo KH, Besra GS, Chatterjee D, Brennan PJ, Belisle JT, et al. The *pimB* gene of *Mycobacterium tuberculosis* encodes a mannosyltransferase involved in lipoarabinomannan biosynthesis. J Biol Chem. 1999;274:31625-31.

[120] Mishra AK, Batt S, Krumbach K, Eggeling L, Besra GS. Characterization of the *Corynebacterium glutamicum* deltapimB' deltamgtA double deletion mutant and the role of *Mycobacterium tuberculosis* orthologues Rv2188c and Rv0557 in glycolipid biosynthesis. J Bacteriol. 2009;191:4465-72.

[121] Tatituri RV, Illarionov PA, Dover LG, Nigou J, Gilleron M, Hitchen P, et al. Inactivation of *Corynebacterium glutamicum* NCgl0452 and the role of MgtA in the biosynthesis of a novel mannosylated glycolipid involved in lipomannan biosynthesis. J Biol Chem. 2007;282:4561-72.

[122] Batt SM, Jabeen T, Mishra AK, Veerapen N, Krumbach K, Eggeling L, et al. Acceptor substrate discrimination in phosphatidyl-myo-inositol mannoside synthesis: structural and mutational analysis of mannosyltransferase *Corynebacterium glutamicum* PimB'. J Biol Chem. 2010;285:37741-52.

[123] Kordulakova J, Gilleron M, Puzo G, Brennan PJ, Gicquel B, Mikusova K, et al. Identification of the required acyltransferase step in the biosynthesis of the phosphatidylinositol mannosides of *Mycobacterium* species. J Biol Chem. 2003;278:36285-95.

[124] Kremer L, Gurcha SS, Bifani P, Hitchen PG, Baulard A, Morris HR, et al. Characterization of a putative alpha-mannosyltransferase involved in phosphatidylinositol trimannoside biosynthesis in *Mycobacterium tuberculosis*. Biochem J. 2002;363:437-47.

[125] Besra GS, Morehouse CB, Rittner CM, Waechter CJ, Brennan PJ. Biosynthesis of mycobacterial lipoarabinomannan. J Biol Chem. 1997;272:18460-6.

[126] Gurcha SS, Baulard AR, Kremer L, Locht C, Moody DB, Muhlecker W, et al. Ppm1, a novel polyprenol monophosphomannose synthase from *Mycobacterium tuberculosis*. Biochem J. 2002;365:441-50.

[127] Gibson KJ, Eggeling L, Maughan WN, Krumbach K, Gurcha SS, Nigou J, et al. Disruption of Cg-Ppm1, a polyprenyl monophosphomannose synthase, and the generation of lipoglycan-less mutants in *Corynebacterium glutamicum*. J Biol Chem. 2003;278:40842-50.

[128] Morita YS, Patterson JH, Billman-Jacobe H, McConville MJ. Biosynthesis of mycobacterial phosphatidylinositol mannosides. Biochem J. 2004;378:589-97.

[129] Kovacevic S, Anderson D, Morita YS, Patterson J, Haites R, McMillan BN, et al. Identification of a novel protein with a role in lipoarabinomannan biosynthesis in mycobacteria. J Biol Chem. 2006;281:9011-7.

[130] Crellin PK, Kovacevic S, Martin KL, Brammananth R, Morita YS, Billman-Jacobe H, et al. Mutations in *pimE* restore lipoarabinomannan synthesis and growth in a *Mycobacterium smegmatis lpqW* mutant. J Bacteriol. 2008;190:3690-9.

[131] Marland Z, Beddoe T, Zaker-Tabrizi L, Lucet IS, Brammananth R, Whisstock JC, et al. Hijacking of a substrate-binding protein scaffold for use in mycobacterial cell wall biosynthesis. J Mol Biol. 2006;359:983-97.

[132] Besra GS, Brennan PJ. The mycobacterial cell wall: biosynthesis of arabinogalactan and lipoarabinomannan. Biochem Soc Trans. 1997;25:845-50.

[133] Mishra AK, Alderwick LJ, Rittmann D, Wang C, Bhatt A, Jacobs WR, Jr., et al. Identification of a novel alpha(1-->6) mannopyranosyltransferase MptB from *Corynebacterium glutamicum* by deletion of a conserved gene, *NCgl1505*, affords a lipomannan- and lipoarabinomannan-deficient mutant. Mol Microbiol. 2008;68:1595-613.

[134] Kaur D, Berg S, Dinadayala P, Gicquel B, Chatterjee D, McNeil MR, et al. Biosynthesis of mycobacterial lipoarabinomannan: role of a branching mannosyltransferase. Proc Natl Acad Sci U S A. 2006;103:13664-9.

[135] Kaur D, McNeil MR, Khoo KH, Chatterjee D, Crick DC, Jackson M, et al. New insights into the biosynthesis of mycobacterial lipomannan arising from deletion of a conserved gene. J Biol Chem. 2007;282:27133-40.

[136] Mishra AK, Alderwick LJ, Rittmann D, Tatituri RV, Nigou J, Gilleron M, et al. Identification of an alpha(1-->6) mannopyranosyltransferase (MptA), involved in *Corynebacterium glutamicum* lipomannan biosynthesis, and identification of its orthologue in *Mycobacterium tuberculosis*. Mol Microbiol. 2007;65:1503-17.

[137] Kaur D, Obregon-Henao A, Pham H, Chatterjee D, Brennan PJ, Jackson M. Lipoarabinomannan of *Mycobacterium*: mannose capping by a multifunctional terminal mannosyltransferase. Proc Natl Acad Sci U S A. 2008;105:17973-7.

[138] Sena CB, Fukuda T, Miyanagi K, Matsumoto S, Kobayashi K, Murakami Y, et al. Controlled expression of branch-forming mannosyltransferase is critical for mycobacterial lipoarabinomannan biosynthesis. J Biol Chem. 2010;285:13326-36.

[139] Birch HL, Alderwick LJ, Bhatt A, Rittmann D, Krumbach K, Singh A, et al. Biosynthesis of mycobacterial arabinogalactan: identification of a novel alpha(1-->3) arabinofuranosyltransferase. Mol Microbiol. 2008;69:1191-206.

[140] Alderwick LJ, Lloyd GS, Ghadbane H, May JW, Bhatt A, Eggeling L, et al. The C-terminal domain of the arabinosyltransferase *Mycobacterium tuberculosis* EmbC is a lectin-like carbohydrate binding module. PLoS Pathog. 2011;7:e1001299.

[141] Skovierova H, Larrouy-Maumus G, Zhang J, Kaur D, Barilone N, Kordulakova J, et al. AftD, a novel essential arabinofuranosyltransferase from mycobacteria. Glycobiology. 2009;19:1235-47.

[142] Zhang N, Torrelles JB, McNeil MR, Escuyer VE, Khoo KH, Brennan PJ, et al. The Emb proteins of mycobacteria direct arabinosylation of lipoarabinomannan and arabinogalactan via an N-terminal recognition region and a C-terminal synthetic region. Mol Microbiol. 2003;50:69-76.

[143] Chatterjee D, Hunter SW, McNeil M, Brennan PJ. Lipoarabinomannan. Multiglycosylated form of the mycobacterial mannosylphosphatidylinositols. J Biol Chem. 1992;267:6228-33.

[144] Appelmelk BJ, den Dunnen J, Driessen NN, Ummels R, Pak M, Nigou J, et al. The mannose cap of mycobacterial lipoarabinomannan does not dominate the *Mycobacterium*-host interaction. Cell Microbiol. 2008;10:930-44.

[145] Khoo KH, Dell A, Morris HR, Brennan PJ, Chatterjee D. Inositol phosphate capping of the nonreducing termini of lipoarabinomannan from rapidly growing strains of *Mycobacterium*. J Biol Chem. 1995;270:12380-9.

[146] Mishra AK, Klein C, Gurcha SS, Alderwick LJ, Babu P, Hitchen PG, et al. Structural characterization and functional properties of a novel lipomannan variant isolated from a *Corynebacterium glutamicum* pimB' mutant. Antonie Van Leeuwenhoek. 2008;94:277-87.

[147] Tatituri RV, Alderwick LJ, Mishra AK, Nigou J, Gilleron M, Krumbach K, et al. Structural characterization of a partially arabinosylated lipoarabinomannan variant isolated from a *Corynebacterium glutamicum* ubiA mutant. Microbiology. 2007;153:2621-9.

[148] Wolucka BA, McNeil MR, de Hoffmann E, Chojnacki T, Brennan PJ. Recognition of the lipid intermediate for arabinogalactan/arabinomannan biosynthesis and its relation to the mode of action of ethambutol on mycobacteria. J Biol Chem. 1994;269:23328-35.

[149] Takayama K, Schnoes HK, Semmler EJ. Characterization of the alkali-stable mannophospholipids of *Mycobacterium smegmatis*. Biochim Biophys Acta. 1973;316:212-21.

[150] Bailey AM, Mahapatra S, Brennan PJ, Crick DC. Identification, cloning, purification, and enzymatic characterization of *Mycobacterium tuberculosis* 1-deoxy-D-xylulose 5-phosphate synthase. Glycobiology. 2002;12:813-20.

[151] Dhiman RK, Schaeffer ML, Bailey AM, Testa CA, Scherman H, Crick DC. 1- Deoxy-D-xylulose 5-phosphate reductoisomerase (IspC) from *Mycobacterium tuberculosis*: towards understanding mycobacterial resistance to fosmidomycin. J Bacteriol. 2005;187:8395-402.

[152] Argyrou A, Blanchard JS. Kinetic and chemical mechanism of *Mycobacterium tuberculosis* 1-deoxy-deoxy-D-xylulose -5-phosphate isomeroreductase. Biochemistry. 2004;43:4375-84.

[153] Takayama K, Goldman DS. Enzymatic synthesis of mannosyl-1-phosphoryl-decaprenol by a cell-free system of *Mycobacterium tuberculosis*. J Biol Chem. 1970;245:6251-7.

[154] Wolucka BA, de Hoffmann E. The presence of beta-D-ribosyl-1-monophosphodecaprenol in mycobacteria. J Biol Chem. 1995;270:20151-5.

[155] Wolucka BA, de Hoffmann E. Isolation and characterization of the major form of polyprenyl-phospho-mannose from *Mycobacterium smegmatis*. Glycobiology. 1998;8:955-62.

[156] Baulard AR, Gurcha SS, Engohang-Ndong J, Gouffi K, Locht C, Besra GS. *In vivo* interaction between the polyprenol phosphate mannose synthase Ppm1 and the integral membrane protein Ppm2 from *Mycobacterium smegmatis* revealed by a bacterial two-hybrid system. J Biol Chem. 2003;278:2242-8.

[157] Scherman H, Kaur D, Pham H, Skovierova H, Jackson M, Brennan PJ. Identification of a polyprenylphosphomannosyl synthase involved in the synthesis of mycobacterial mannosides. J Bacteriol. 2009;191:6769-72.

[158] Skovierova H, Larrouy-Maumus G, Pham H, Belanova M, Barilone N, Dasgupta A, et al. Biosynthetic origin of the galactosamine substituent of arabinogalactan in *Mycobacterium tuberculosis*. J Biol Chem. 2010;285:41348-55.

[159] Klutts JS, Hatanaka K, Pan YT, Elbein AD. Biosynthesis of D-arabinose in *Mycobacterium smegmatis*: specific labeling from D-glucose. Arch Biochem Biophys. 2002;398:229-39.

[160] Scherman M, Weston A, Duncan K, Whittington A, Upton R, Deng L, et al. Biosynthetic origin of mycobacterial cell wall arabinosyl residues. J Bacteriol. 1995;177:7125-30.

[161] Scherman MS, Kalbe-Bournonville L, Bush D, Xin Y, Deng L, McNeil M. Polyprenylphosphate-pentoses in mycobacteria are synthesized from 5-phosphoribose pyrophosphate. J Biol Chem. 1996;271:29652-8.

[162] Huang H, Scherman MS, D'Haeze W, Vereecke D, Holsters M, Crick DC, et al. Identification and active expression of the *Mycobacterium tuberculosis* gene encoding 5-phospho-D-ribose-1-diphosphate: decaprenyl-phosphate 5-phosphoribosyltransferase, the first enzyme committed to decaprenylphosphoryl-d-arabinose synthesis. J Biol Chem. 2005;280:24539-43.

[163] Mikusova K, Huang H, Yagi T, Holsters M, Vereecke D, D'Haeze W, et al. Decaprenylphosphoryl arabinofuranose, the donor of the D-arabinofuranosyl residues of mycobacterial arabinan, is formed via a two-step epimerization of decaprenylphosphoryl ribose. J Bacteriol. 2005;187:8020-5.

[164] Christophe T, Jackson M, Jeon HK, Fenistein D, Contreras-Dominguez M, Kim J, et al. High content screening identifies decaprenyl-phosphoribose 2' epimerase as a target for intracellular antimycobacterial inhibitors. PLoS Pathog. 2009;5:e1000645.

[165] Makarov V, Manina G, Mikusova K, Mollmann U, Ryabova O, Saint-Joanis B, et al. Benzothiazinones kill *Mycobacterium tuberculosis* by blocking arabinan synthesis. Science. 2009;324:801-4.

[166] Makarov V, Riabova OB, Yuschenko A, Urlyapova N, Daudova A, Zipfel PF, et al. Synthesis and antileprosy activity of some dialkyldithiocarbamates. J Antimicrob Chemother. 2006;57:1134-8.

[167] Pasca MR, Degiacomi G, Ribeiro AL, Zara F, De Mori P, Heym B, et al. Clinical isolates of *Mycobacterium tuberculosis* in four European hospitals are uniformly susceptible to benzothiazinones. Antimicrob Agents Chemother. 2010;54:1616-8.

[168] Crellin PK, Brammananth R, Coppel RL. Decaprenylphosphoryl-beta- D-ribose 2'-epimerase, the target of benzothiazinones and dinitrobenzamides, is an essential enzyme in *Mycobacterium smegmatis*. PLoS One. 2011;6:e16869.

[169] Manina G, Pasca MR, Buroni S, De Rossi E, Riccardi G. Decaprenylphosphoryl-beta- D-ribose 2'-epimerase from *Mycobacterium tuberculosis* is a magic drug target. Curr Med Chem. 2010;17:3099-108.

[170] Batt SM, Jabeen T, Bhowruth V, Quill L, Lund PA, Eggeling L, et al. Structural basis of inhibition of *Mycobacterium tuberculosis* DprE1 by benzothiazinone inhibitors. Proc Natl Acad Sci U S A. 2012;109:11354-9.

[171] Tarnok I, Tarnok Z. Carotenes and xanthophylls in mycobacteria. II. Lycopene, alpha- and beta-carotene and xanthophyll in mycobacterial pigments. Tubercle. 1971;52:127-35.

[172] Houssaini-Iraqui M, Lazraq MH, Clavel-Seres S, Rastogi N, David HL. Cloning and expression of *Mycobacterium aurum* carotenogenesis genes in *Mycobacterium smegmatis*. FEMS Microbiol Lett. 1992;69:239-44.

[173] Ramakrishnan L, Tran HT, Federspiel NA, Falkow S. A crtB homolog essential for photochromogenicity in *Mycobacterium marinum*: isolation, characterization, and gene disruption via homologous recombination. J Bacteriol. 1997;179:5862-8.

[174] Viveiros M, Krubasik P, Sandmann G, Houssaini-Iraqui M. Structural and functional analysis of the gene cluster encoding carotenoid biosynthesis in *Mycobacterium aurum* A+. FEMS Microbiol Lett. 2000;187:95-101.

[175] Robledo JA, Murillo AM, Rouzaud F. Physiological role and potential clinical interest of mycobacterial pigments. IUBMB Life. 2011;63:71-8.

[176] Kim SW, Keasling JD. Metabolic engineering of the nonmevalonate isopentenyl diphosphate synthesis pathway in *Escherichia coli* enhances lycopene production. Biotechnol Bioeng. 2001;72:408-15.

[177] Ito M, Yamano Y, Tode C, Wada A. Carotenoid synthesis: retrospect and recent progress. Arch Biochem Biophys. 2009;483:224-8.

[178] Misawa N, Satomi Y, Kondo K, Yokoyama A, Kajiwara S, Saito T, et al. Structure and functional analysis of a marine bacterial carotenoid biosynthesis gene cluster and astaxanthin biosynthetic pathway proposed at the gene level. J Bacteriol. 1995;177:6575-84.

[179] Wenk MR. The emerging field of lipidomics. Nat Rev Drug Discov. 2005;4:594-610.

[180] Wenk MR. Lipidomics: new tools and applications. Cell. 2010;143:888-95.

[181] Layre E, Sweet L, Hong S, Madigan CA, Desjardins D, Young DC, et al. A comparative lipidomics platform for chemotaxonomic analysis of *Mycobacterium tuberculosis*. Chem Biol. 2011;18:1537-49.

[182] Sartain MJ, Dick DL, Rithner CD, Crick DC, Belisle JT. Lipidomic analyses of *Mycobacterium tuberculosis* based on accurate mass measurements and the novel "Mtb LipidDB". J Lipid Res. 2011;52:861-72.

[183] Madigan CA, Cheng TY, Layre E, Young DC, McConnell MJ, Debono CA, et al. Lipidomic discovery of deoxysiderophores reveals a revised mycobactin biosynthesis pathway in *Mycobacterium tuberculosis*. Proc Natl Acad Sci U S A. 2012;109:1257-62.

Permissions

The contributors of this book come from diverse backgrounds, making this book a truly international effort. This book will bring forth new frontiers with its revolutionizing research information and detailed analysis of the nascent developments around the world.

We would like to thank Rodrigo Valenzuela Baez, PhD, for lending his expertise to make the book truly unique. He has played a crucial role in the development of this book. Without his invaluable contribution this book wouldn't have been possible. He has made vital efforts to compile up to date information on the varied aspects of this subject to make this book a valuable addition to the collection of many professionals and students.

This book was conceptualized with the vision of imparting up-to-date information and advanced data in this field. To ensure the same, a matchless editorial board was set up. Every individual on the board went through rigorous rounds of assessment to prove their worth. After which they invested a large part of their time researching and compiling the most relevant data for our readers. Conferences and sessions were held from time to time between the editorial board and the contributing authors to present the data in the most comprehensible form. The editorial team has worked tirelessly to provide valuable and valid information to help people across the globe.

Every chapter published in this book has been scrutinized by our experts. Their significance has been extensively debated. The topics covered herein carry significant findings which will fuel the growth of the discipline. They may even be implemented as practical applications or may be referred to as a beginning point for another development. Chapters in this book were first published by InTech; hereby published with permission under the Creative Commons Attribution License or equivalent.

The editorial board has been involved in producing this book since its inception. They have spent rigorous hours researching and exploring the diverse topics which have resulted in the successful publishing of this book. They have passed on their knowledge of decades through this book. To expedite this challenging task, the publisher supported the team at every step. A small team of assistant editors was also appointed to further simplify the editing procedure and attain best results for the readers.

Our editorial team has been hand-picked from every corner of the world. Their multi-ethnicity adds dynamic inputs to the discussions which result in innovative outcomes. These outcomes are then further discussed with the researchers and contributors who give their valuable feedback and opinion regarding the same. The feedback is then collaborated with the researches and they are edited in a comprehensive manner to aid the understanding of the subject.

Apart from the editorial board, the designing team has also invested a significant amount of their time in understanding the subject and creating the most relevant covers. They scrutinized every image to scout for the most suitable representation of the subject and create an appropriate cover for the book.

The publishing team has been involved in this book since its early stages. They were actively engaged in every process, be it collecting the data, connecting with the contributors or procuring relevant information. The team has been an ardent support to the editorial, designing and production team. Their endless efforts to recruit the best for this project, has resulted in the accomplishment of this book. They are a veteran in the field of academics and their pool of knowledge is as vast as their experience in printing. Their expertise and guidance has proved useful at every step. Their uncompromising quality standards have made this book an exceptional effort. Their encouragement from time to time has been an inspiration for everyone.

The publisher and the editorial board hope that this book will prove to be a valuable piece of knowledge for researchers, students, practitioners and scholars across the globe.

List of Contributors

Rodrigo Valenzuela B.
Nutrition and Dietetics School, Faculty of Medicine, University of Chile, Santiago Chile

Alfonso Valenzuela B.
Lipid Center, Nutrition and Food Technology Institute, University of Chile, Faculty of Medicine, University of Los Andes, Santiago Chile

Claudia Borza, Danina Muntean, Cristina Dehelean, Germaine Săvoiu, Corina Şerban, Georgeta Simu, Mihaiela Andoni, Marius Butur and Simona Drăgan
University of Medicine and Pharmacy "Victor Babeş" Timişoara, Romania

Line M. Grønning-Wang, Christian Bindesbøll and Hilde I. Nebb
The Medical Faculty, Institute of Basic Medical Sciences, Department of Nutrition, University of Oslo, Norway

Jason L. Burkhead
Department of Biological Sciences, University of Alaska Anchorage, Anchorage, AK, USA

Svetlana Lutsenko
Department of Physiology, Johns Hopkins University, Baltimore, MD, USA

Jasmina Dimitrova-Shumkovska
Institute of Biology, Department of Experimental Biochemistry and Physiology, Faculty of Natural Sciences and Mathematics, Ss. Cyril and Methodius University - Skopje, Republic of Macedonia

Leo Veenman, Inbar Roim and Moshe Gavish
Rappaport Family Institute for Research in the Medical Sciences, Technion-Israel Institute of Technology, Department of Molecular Pharmacology, Haifa, Israel

Yasuo Uchiyama and Eiki Kominami
Departments of Cell Biology and Neurosciences, and Biochemistry, Juntendo University Graduate School of Medicine, Bunkyo-ku, Tokyo, Japan

Alicia Huazano-García and Mercedes G. López
Departamento de Biotecnología y Bioquímica, Centro de Investigación y de Estudios Avanzados del IPN, México

Paul K. Crellin
Australian Research Council Centre of Excellence in Structural and Functional Microbial Genomics, Department of Microbiology, Monash University, Australia

Chu-Yuan Luo and Yasu S. Morita
Department of Microbiology, University of Massachusetts, Amherst, USA

Printed in the USA
CPSIA information can be obtained
at www.ICGtesting.com
JSHW011356221024
72173JS00003B/301